PRISONED CHICKENS, POISONED EGGS

An Inside Look at the Modern Poultry Industry

Revised Edition

Karen Davis, PhD

Book Publishing Company

Summertown, Tennessee

Cover Design: Warren Jefferson
Interior Design: Cord Slatton
Edited by: Cheryl Redmond

Published by
Book Publishing Company
P.O. Box 99
Summertown, TN 38483
1-888-260-8458
www.bookpubco.com

Printed in Canada

ISBN: 978-1-57067-229-3

15 14 13 12 11 10 8 7 6 5 4 3

Library of Congress Cataloging-in-Publication data
Davis, Karen, 1944-
Prisoned chickens, poisoned eggs : an inside look at the modern poultry industry / by Karen Davis. -- [2nd ed.].
p. cm.
Includes bibliographical references and index.
ISBN 978-1-57067-229-3
1. Chickens. 2. Chickens--Diseases. 3. Eggs--Production. 4. Chicken industry. 5. Egg trade. I. Title.

SF487.D27 2009
636.5'0896--dc22

2008012249

Book Publishing Company is a member of Green Press Initiative. We chose to print this title on paper with postconsumer recycled content, processed without chlorine, which saved the following natural resources:

5 trees
2 million BTU of total energy
493 lbs. of greenhouse gases
2,375 gallons of water
144 lbs. of solid waste

For more information, visit www.greenpressinitiative.org.
Paper calculations from the Environmental Defense Paper Calculator: www.edf.org/papercalculator

TABLE OF CONTENTS

Preface to the New Edition

I wrote *Prisoned Chickens, Poisoned Eggs* in the mid-1990s in order to bring attention to the billions of chickens buried alive on factory farms. At the time, neither the animal rights movement nor the public at large knew very much about chickens or about how the poultry industry originated and developed in twentieth-century America to become the model for industrialized farmed-animal production around the world. Some informative articles and book chapters had appeared, but the poultry industry's own detailed and glowing account of its transformation of the chicken, from an active outdoor bird scouring the woods and fields to a sedentary indoor meat-and-egg "machine," filled with suffering, diseases, and antibiotics, remained largely unknown.

The purpose of *Prisoned Chickens, Poisoned Eggs* was to bring this story to light in a way that would reveal the tragedy of chickens through the lens of the industry that created their tragedy without pity or guilt. The book became, as I'd hoped it would, a blueprint for people seeking a coherent picture of the U.S. poultry industry, as well as a handbook for animal rights activists seeking to develop effective strategies to expose and relieve the plight of chickens.

While much has happened since *Prisoned Chickens, Poisoned Eggs* first appeared in 1996, little has changed for the chickens themselves, except that their lives have become, as a global phenomenon, even more miserable. Instead of 7.5 billion chickens being slaughtered in the mid-1990s in the United States, nearly 10 billion chickens are now being slaughtered, with parallel rises in other countries reflecting the expansion of chicken consumption and industrialized production into Latin America, China, India, Africa, Russia, Mexico, and elsewhere. Throughout the world, over 40 billion chickens are now being slaughtered for meat each year, and over 5 billion hens are in battery cages, many of them in egg-production complexes holding up to a million or more birds.

U.S. per capita consumption of chicken went from 42 pounds in 1972 to 48.4 pounds in 1995 to 86 pounds in 2007, and the number of "meat-type" chickens being crammed into filthy, dark sheds has risen from 20,000 up to as many as 50,000 birds in a single metal building. Chickens forced to gain five pounds in six weeks in the mid-1990s are now being forced to gain eight pounds in the same amount of time. Chickens raised for meat are in such crippling pain, their bodies are so abnormally large and disproportioned, their skin, joints, and intestines are so rotting with ulcers and necrotic diseases, that they can barely stand, let alone walk—not that there is anywhere for them to go, being raised as they are without room to move except onto an adjacent bird's back. When you pick up a chicken on the road who has fallen off a truck on the way to the slaughter plant, the huge white bird with the little peeping voice and baby blue eyes feels like liquid cement.

When I decided to start an advocacy group for chickens in the late 1980s, I was told by some that people weren't "ready" for chickens. This proved to be false. The point, in any case, was to *make* people ready. Between 1996, when *Prisoned Chickens, Poisoned Eggs* first appeared, and the start of the present century, up to the time of this writing in 2008, chickens have become a primary focus of the animal advocacy movement. United Poultry Concerns' campaign in the 1990s to expose the U.S. egg-industry's practice of starving hens for weeks at a time, to force them to molt their feathers and reduce the cost of egg production, had a decisive effect in shifting attention toward chickens.

Most notably, activists in the United States and Canada have joined with their European and Australian allies in waging a high-profile campaign to get rid of the barren battery cage system for laying hens, though the European ban set for 2012 is not without problems, as I show in chapter 3. Farmed-animal sanctuaries and stories about rescued chickens have become an important part of the animal advocacy movement, along with shocking, well-publicized, undercover investigations documenting the appalling cruelty to chickens and turkeys by poultry industry workers.

Recently, a number of celebrity chefs, environmentalists, and sectors of the food industry have come on board with animal welfare groups calling for an end to industrialized poultry and egg production in favor of "humane," "free-range" methods of production. These alliances have spawned concern and anxiety among those animal advocates who consider it a betrayal to work for welfare reforms that involve the duplicity of encouraging people to eat animals and animal products under the illusion that the animals involved were somehow raised and butchered "humanely."

Instead, it is argued that animal advocacy programs should be mainly directed at getting people off animal products altogether. Farmed-animal advocates should invest their resources in campaigns designed to increase the demand for vegan foods—the delicious mock meats, soymilks, egg substitutes, and other vegan items that are now readily available.

The following chapters tell the story of chickens and the poultry industry and where things stand as we move toward the end of the first decade of the new century in a world in which avian influenza, food poisoning, global warming, vegan cuisine, genetic engineering, animal welfare campaigns, and the expansion of poultry and egg production and consumption jostle together. At the heart of this story is the chicken, to whom this book is dedicated, and on whose side, and at whose side, I remain steadfast.

Karen Davis, PhD, President
United Poultry Concerns
PO Box 150
Machipongo, VA 23405
(757) 678-7875
Karen@upc-online.org

ACKNOWLEDGEMENTS

I'm surrounded by file cabinets bulging with manila folders that are filled with the thousands of articles I have accumulated, read, and re-read over a 20-year period about chickens, turkeys, ducks, and other domestic fowl in the areas of food production, science, education, entertainment, and human companionship situations. Most of these articles were photocopied from the stacks of agribusiness periodicals I began subscribing to in the late 1980s and continue to read and accumulate along with the storm of documents that now appear on the Internet. I therefore acknowledge the assistance of these disheartening but indispensable publications in bringing the plight of farmed animals and birds in particular to the light that I have chosen to cast upon them in these pages.

Hundreds of articles about the poultry and egg industry have been brought to my attention by Mary Finelli, starting when she was a researcher in the farm animal division of The Humane Society of the United States and continuing to the present time in her capacity as editor of the online news digest Farmed Animal Watch.

In addition, I am grateful to Clare Druce, my mentor and friend and cofounder of the pressure group Chickens' Lib, which began campaigning for the abolition of cruel methods of poultry-keeping in the United Kingdom in the early 1970s, now called Farm Animal Welfare Network. Clare's 1989 book *Chicken and Egg: Who Pays the Price?* was the first, and to my knowledge the only other book besides my own, to have dealt specifically with industrialized poultry and egg production from an animal advocacy point of view.

I also wish to thank Patty Mark, founder of Animal Liberation Victoria in Australia. Patty has worked for 30 years on behalf of chickens and other farmed animals. Her personal investigations of factory farms led her to organize, in 1993, the first Open Rescue in Australia. Now a worldwide method of documenting the hidden atrocities of factory farming, it was Patty's presentation at United

Poultry Concerns' 1999 conference on direct action for animals, which pioneered the Open Rescue in North America. This form of undercover investigation laid the groundwork for the campaigns that have since followed to educate the public about the horrible treatment of animals raised and slaughtered for food in the United States and Canada.

The horrible treatment of chickens was laid out in excruciating detail by former Tyson employee Virgil Butler, to whom chapter five of this book on the death of chickens at slaughter owes much. Virgil's revelations in 2003 led to my incessant questioning of him about the chicken slaughter process and culture. His e-mails are contained in a three-ring notebook from which I have extracted passages for this book. Virgil gave a riveting talk at United Poultry Concerns' 2004 conference in Norfolk, Virginia, that is fortunately preserved on DVD. Virgil died on December 15, 2006. To his authentic voice for chickens and a better world, I am beholden.

Finally, it is my privilege to acknowledge and thank the supporters of United Poultry Concerns around the world, including the editors, publishers, and artists of the Book Publishing Company who have made *Prisoned Chickens, Poisoned Eggs* possible. In doing so, I draw attention to the fact that millions of people care deeply about chickens and turkeys and other birds. Through the years, countless people have inspired and educated me with their beautiful stories and photographs of the birds they love. Together with the birds themselves, they are the kindred spirits who keep me going, and I can never thank them enough.

PROLOGUE

He woke up on the floor of the broiler shed with 30 thousand other bewildered young chickens under the electric lights, with the familiar pain in his throat and a burning sensation deep inside his eyes. . . . He saw green leaves shining through flashes of sunlight, as he peeked through his mother's feathers and heard the soft, awakening cheeps of his brothers and sisters, and felt his mother's heart beating next to his own through her big, warm body surrounding him, which was his world.

A crow had cried out, and another cried out again.

He started—the spry, young jungle fowl was ready for the day, ready to begin scratching the soil, which he had known by heart ever since way back when chickenhood first arose in the tropical magic mornings of the early world. In the jungle forest, the delicious seeds of bamboo that are hidden beneath the leaves on the ground are treasured in the heart of the chicken.

The rooster called out excitedly: "Family, come see what food I've found for you this morning!" . . .

His aching legs—they brought him back to reality as he closed his eyes stinging with ammonia burn—could not move. They could no longer bear the weight of flesh that bore down upon them, which was definitely not the body of a mother hen. A mother hen, an ancestral memory kept telling him over and over, had once shushed and lulled him to sleep, pressed against her body nestled deep inside her wings, which fluffed over him when he was a chick. That was a long time ago, long before he was a "broiler" chicken, crippled and encased in these cells of fat and skeletal pain. He was turning purple. His lungs filled slowly with fluid, leaking from his vessels backward through the valves of his heart, as he stretched out on the filthy litter in a final spasm of agony, and died.

—Karen Davis, "Memories Inside a Broiler Chicken House"

I remember how wonderful it was to peck my way through the shell and step out into the warm, bright dawn of life. I have seen no other sunrise. We live in eternal noontime. My birth was a grievous mistake. And yet an egg is developing in me, as always. I can't stop it. I feel its growth, and despite all my bitterness, tiny surges of tenderness fill me. How I wish I could stop the egg from growing so that I wouldn't have to know these tender feelings. But I can't stop. I'm an egg machine, the best egg machine in the world.

"Don't be so gloomy, Sister. There are better times coming."

The insane hen in the cage beside mine has fallen victim to a common delusion here at the egg factory. "No better times are coming, Sister," I reply. "Only worse times."

"You're mistaken, my dear. I happen to know. Very soon we'll be scratching in a lovely yard."

I don't bother to reply. She's cheered by her delusions. And since our end will be the same, what does it matter how we spend our days here? Let her dream in her lovely yard. Let her develop her dream, to its fullest, until she imagines that the wire floor beneath her claws has become warm, dry earth. We don't have much longer to go. Our life span is only fourteen months of egg laying and then we're through.

An egg machine!

There's a great fluttering of wings along this cell block, and much loud clucking. The cages are opening, and one by one rough hands grab us.

"You see, Sister. I told you better times were coming. Now we're finally going."

Now we're hung upside down, our feet are tied together with wire. "You see, Sister. It's just as I told you—the better times have come at last."

We're hooked to a slowly moving belt. Hanging upside down, we're carried along through a dark tunnel. The wire bites into my flesh. Swaying through the darkness we go. The gurgling cries up ahead of us make clear what better times have come.

"Our reward, Sister, is here at last," cries our mad sister.

"We were good and laid many eggs and now we get our reward."

The cry of each hen is cut off so that her squawking becomes liquid bubbling. And then the sound of dripping: drip, drip, drip.

"Oh, I can see it now, Sister, the lovely yard I spoke of, all covered with red flowers and . . ."

The mash runs out of her neck.

—William Kotzwinkle, *Doctor Rat*

INTRODUCTION

I did not grow up around chickens. As is probably the case for most people growing up in post-World War II America, my personal acquaintance with chickens and other animals on the farm was confined to experiences at the table. There were some brief encounters with baby "Easter chickens" and rabbits way back in childhood, and a long-suppressed witnessing of a brown hen beheaded on a chopping block with an axe by a playmate's father.

However, a chicken named Viva changed the course of my life and career. When I met her, I was an English teacher completing my doctoral dissertation in English at the University of Maryland. I had expected to teach English for the rest of my life. Yet during the mid- to late 1980s, I found myself increasingly drawn to the plight of nonhuman animals in human society, particularly farmed animals. The huge number of factory-farmed animals was astonishing. At the bottom of this pile were billions of birds imprisoned in intensive confinement systems, totally out of sight. Farmed animals were generally dismissed as beyond the pale of equal, or even any, moral concern because, it was argued, they had been bred to a substandard state of intelligence and biological fitness, and because they were "just food" that was "going to be killed anyway."

My experience with Viva, a crippled and abandoned "broiler" hen, put these matters into perspective. Viva was expressive, responsive, communicative, affectionate, and alert. Though she was cursed with a manmade body, there was nothing inferior about her personally. She already had a voice, but her voice needed to be amplified within the oppressive human system in which she was trapped. There were billions of Vivas out there, just as special.

Viva's death was painful, but knowing her clarified my future. It was not only that Viva had suffered, but that she was a valuable being, somebody worth fighting for. She was not "just a chicken." She was a chicken. She was a member of earth's community, a dignified being with a claim equal to anyone else's to justice, compassion, and a life worth living.

This book is dedicated to her and to the making of a future in which every Viva has a voice that is heard.

CHAPTER 1

HISTORY

It is a far cry from the time that man first heard the crow of the wild cock of the bamboo jungles of India to the cackle of the highly domesticated hen upon celebrating her production of 1,000 or more eggs.

—M. A. Jull, "The Races of Domestic Fowl," *National Geographic*, April 1927, 379

It would be rash to suggest that before the twentieth century the life of chickens was rosy. In the eighteenth century, the New Jersey Quaker John Woolman noted the despondency of chickens on a boat going from America to England and the poignancy of their hopeful response when they came close to land. Behind them lay centuries of domestication, preceded and paralleled by an autonomous life in the tropical forests of Southeast Asia that persists to this day. Ahead lay a fate that premonition would have tried in vain to prevent from coming to pass. This book is about that fate, the fate of chickens in our society.

Chickens are creatures of the earth who no longer live on the land. In the industrialized world, billions of chickens are locked inside factory-farm buildings, and billions more are similarly confined in Africa, Asia, India, China, and other parts of the world where poultry factory farming is rapidly supplanting traditional small farming. If there is such a thing as "earthrights," the right of a creature to experience directly the earth from which it was derived and on which its happiness in life chiefly depends, then chickens have been stripped of theirs. They have not changed; however, the world in which they live has been disrupted for human convenience against their will.

Early History

The Red Jungle Fowl, generally thought to be the primary ancestor of the domestic chicken, is found in the foothills of the Himalayan Mountains in the north of India to tropical Southeast Asia. . . . Jungle fowl live in a variety of forested habitats.

—Bell and Weaver, *Commercial Chicken Meat and Egg Production*, 72

People have kept chickens for thousands of years, probably beginning in Southeast Asia, where it is speculated that one or more species of jungle fowl contributed to the modern domesticated fowl with the possible involvement of other wild birds, such as the grouse. It may be that cockfighting preceded and led to the use of chickens for food, with the female birds being perceived as a source of meat and eggs. Humans may have discovered that by stealing from the nest eggs they did not want to hatch, or wanted to eat, they could induce the hen to lay compensatory eggs and continue to lay through an extended season.

The breeding of hens to encourage egg laying may have begun as long as five or even ten thousand years ago. Human intervention is certain. Egg laying as an independent activity detached from the giving of life is not a natural phenomenon in birds. As *The Chicken Book* states, "The chief distinction between domestic and wild fowl lies in the fact that wild fowl (like all wild birds) do not lay a surplus of eggs. Most commonly they lay only in the spring when they are ready to raise a brood of chicks" (Smith and Daniel, 33).

The spread of the jungle fowl from the Indian subcontinent westward to the Mediterranean basin, northern Europe, and Africa, and eastward from China to the Pacific Islands, probably occurred through military and commercial activity. By the fourth century BC, chickens were established in Persia, Greece, and Rome. The ancient Chinese bred heavy chickens for meat. In Persia and Greece, the birds were objects of sacrifice. Cockfighting spread from India to Persia, the Pacific Islands, Greece, and Rome. When Julius Caesar arrived in Britain in 55 BC, he found the native Britons already kept fighting cocks for sport. By the late Middle Ages, cockfighting had spread throughout the Roman Empire.

References to chickens have been found in Egyptian records as early as the fourteenth century BC. The cock is evoked in poetical and pictorial images and in a royal accounting of tribute from the East, that reads, "Lo! four birds of this land, which bring forth every day." Egypt is the first nation on record to have mass-produced chickens and eggs similar to modern practice. Some 4,000 years ago, the Egyptians built fire-heated clay brick incubators that could hatch as many as 10,000 chicks at a time (Lewis, 456, 467). In *Factory Farming,* Andrew Johnson cites the Roman writer Varro to show how in the first century BC the Romans maintained specialized chicken farms "with elaborate hen-houses equipped with ladders, high roosts, nests, and reliable trapdoors to keep out foxes and weasels."

These houses accommodated between 40 and 200 birds and, depending on the size, were divided into smaller rooms where cocks and their attached hens could roost separately from other families of birds. Parasites such as mites and lice were controlled by smoke piped from the bakery through the chicken house, and periodic evacuation followed by disinfection of the building was apparently practiced then, as now, to control the diseases that develop through overcrowding.

Johnson dismisses the idea that the pre-factory-farming era was idyllic for chickens and other farmed animals, suggesting, rather, that factory farming is an extension of age-old attitudes and practices in regard to animals raised for food. Recalling Elizabethan England in the sixteenth century, he says, for example, that the modern battery-cage building is "little more than a many thousand times larger replica of the housewife's kitchen hen-coop which might at that date have filled in the unused space under the dresser" (Johnson, 23).

Keith Thomas adds to this picture in *Man and the Natural World,* noting that poultry and game birds "were often fattened in darkness and confinement, sometimes being blinded as well. 'The cock being gelded,' it was explained, 'he is called a capon and is crammed [force-fed] in a coop.' Geese were thought to put on weight if the webs of their feet were nailed to the floor; and it was the custom of some seventeenth-century housewives to cut the legs off living fowl in the belief that it made their flesh more tender." The London poulterers, Thomas writes, "kept thousands of live birds in their cellars and attics." In 1842, Edwin Chadwick found that "fowls were still being reared in town bedrooms" (Thomas, 94–95).

In *A Natural History of the Senses,* Diane Ackerman describes culinary practices that arose in England in the eighteenth century when "bored city dwellers became fascinated by sadism," including the idea that "torturing an animal made its meat healthier and better tasting." One recipe starts out: "Take a red cock that is not too old and beat him to death." Another instructs:

> Take a Goose, or a Duck, or some such lively creature, pull off all her feathers, only the head and neck must be spared: then make a fire round about her, not too close to her, that the smoke do not choke her, and that the fire may not burn her too soon; not too far off, that she may not escape free: within the circle of the fire let there be set small cups and pots of water, wherein salt and honey are mingled; and let there be set also chargers full of sodden Apples, cut into small pieces in the dish. The Goose must be all larded, and basted over with butter: put then fire about her, but do not make too much haste, when as you see her begin to roast; for by walking about and flying here and there, being cooped in by the fire that stops her way out the unwearied Goose is kept in; she will fall to drink the water to quench her thirst, and cool her heart and all her body, and the Apple sauce will make her dung and cleanse and empty her. And when she roasteth, and consumes inwardly, always wet her head and heart with a wet sponge; and when you see her giddy with running, and begin to stumble, her heart wants moisture, and she is roasted enough. Take her up and set her before your guests and she will cry as you cut off any part from her and will be almost eaten up before she be dead: it is mighty pleasant to behold! (Ackerman, 147).

Eighteenth- and nineteenth-century literature offers additional testimony regarding the treatment of chickens and other domestic fowl. In Tobias Smollett's novel *The Expedition of Humphry Clinker,* published in 1771, the Welsh traveler Matthew Bramble complains during a visit to London that "the poultry is all rotten, in consequence of a fever, occasioned by the infamous practice of sewing up the gut, that they may be the sooner fattened in coops, in consequence of this cruel retention." He contrasts the crowded poultry in London with the condition of his own birds in the country "that never knew confinement, but when they were at roost" (Smollett, 119, 121).

In Thomas Hardy's novel *Tess of the d'Urbervilles,* set in the middle of the nineteenth century, the principal character, Tess Durbeyfield, works on a poultry farm on a landed estate where the birds—"Hamburghs, Bantams, Cochins, Brahmas, Dorkings, and such other sorts as were in fashion just then"—are crowded into a cottage formerly inhabited by generations of families: "The rooms wherein dozens of infants had wailed at their nursing now resounded with the tapping of nascent chicks.

Distracted hens in coops occupied spots where formerly stood chairs supporting sedate agriculturalists. The chimney-corner and once blazing hearth were now filled with inverted beehives, in which the hens laid their eggs; while out of doors the plots that each succeeding householder had carefully shaped with his spade were torn by the cocks in wildest fashion" (Hardy, 68–70).

As is still the practice in small towns throughout the world, chickens and other birds were taken to market with their legs tied. Tess's father, an improvident, alcoholic boot-haggler pretending to earn a living, carries around a live hen who is forced to lie with her legs tied under a bar table while he wiles away the time drinking (Hardy, 328). Mark Braunstein has described the sale of a chicken that he watched take place in an Italian town some years ago, during which the buyer "clutched the chicken by the legs, several times unknowingly and uncaringly banged its head against the ground, weighed it while yanking it to and fro, and finally dumped it into her sack. Then she must have forgotten something, pulled the chicken out again, but only halfway, stuck its legs into the railings of a nearby fence, left it dangling undoubtedly with broken legs, and walked away" (Braunstein, 93–94).

In addition to these chronicles, there is evidence in history of human regard for chickens, quite apart from economics. In the 1970s, I read about a man in South America who cried when the Peace Corps converted his traditional household flock into a battery-caged hen house. He wept for his hens and the loss of their friendship despite promises that the new "scientific" method of debeaking them and treating them like machines would one day bring him a Cadillac. Eighteenth-century Europeans traveling in South America noted that the Indian women were so fond of their fowls they would not sell them, much less kill them with their own hands: "So that if a Spaniard . . . offer[s] ever so much money for a fowl, they refuse to part with it" (Ritson, 216–217).

In *Letters from an American Farmer*, a study of American Colonial society published in 1782, St. John de Crevecoeur wrote, "I never see an egg brought on my table but I feel penetrated with the wonderful change it would have undergone but for my gluttony; it might have been a gentle, useful hen leading her chickens with a care and vigilance which speaks shame to many women. A cock perhaps, arrayed with the most majestic plumes, tender to its mate, bold, courageous, endowed with an astonishing instinct, with thoughts, with memory, and every distinguishing characteristic of the reason of man" (Stone, 55).

Molly Ivins tells the story of a Texas woman, Mary Ann Goodnight, who was often left alone on a ranch near the Palo Duro Canyon. "One day in 1877, a cowboy rode into her camp with three chickens in a sack as a present for her. He naturally expected her to cook and eat the fowl, but Goodnight kept them as pets. She wrote in her diary, 'No one can ever know how much company they were'" (Ivins, 167).

A touching example of human love for a chicken is told by the British humanitarian writer Henry Salt concerning an old woman he once met in a roadside cottage in the Lake District "who had for her companion, sitting in an armchair by the fire, a lame hen, named Tetty, whom she had saved and reared from chicken-hood." A few years later, Salt encountered the woman again and inquired about Tetty but learned that she was dead. "Ah, poor Tetty!" the woman said in tears. "She passed away several months ago, quite conscious to the end" (Salt, 130–131).

Beginning of the Modern Factory Farm

Commercial poultry farms are especially successful near large centers of population, where the demand is for a strictly fresh, new-laid egg and fresh-killed poultry. Hundreds of such enterprises are being successfully operated in the Atlantic and Pacific Coast States. The eastern sections produce especially for the New York trade, and the Pacific coast sections, after meeting the demands of the larger Pacific coast cities, ship their eggs to the Atlantic seaboard, where they find a ready market at attractive prices. The production of eggs under these conditions is rapidly assuming factory proportions.

–Harry R. Lewis, "America's Debt to the Hen,"
National Geographic, April, 1927, 453

Chickens were the first farmed animals to be permanently confined indoors in large numbers in automated systems based on intensive genetic selection for food-production traits and reliance on antibiotics (substances produced by microorganisms) and other drugs (chemicals). In the twentieth century, the poultry industry in the United States became the model for animal agriculture throughout the world. In India, where half of the population is vegetarian and the majority of people are Hindu, a religion that prohibits or discourages the eating of anything that is or has the potential to be animal life, people have been pressured to adopt intensive poultry production and to consume the unfertilized eggs of hens kept in battery cages. According to a 2005 Worldwatch Institute report, "India now ranks fifth in the world in both broiler and egg production," much of which "is occurring in large factory farms near densely populated cities" (Nierenberg, 48).

Chickens were brought to America by the Europeans. The centennial issue of *Broiler Industry* magazine noted in July 1976, "Nearly every boatload of settlers that came to the New World in the seventeenth and eighteenth centuries brought with it at least a few chickens" (94). In the early nineteenth century, chickens, turkeys, ducks, and geese roamed largely at will, often sharing the farmhouse with their owners. They foraged in the fields and among the bushes and willows of the brooks and springs, and frequented the Colonial dung hills and ash heaps to obtain the grasses, seeds, sprouts, insects, vitamins, and minerals they needed, with little or no dependence on homegrown grain.

Chickens were raised in towns and villages as well as on farms, and many city people kept them in back lots of various sizes. As late as 1830, the average number of chickens for the three million reporting farms in the United States was 23 (Skinner, 218). At that time, many chickens still seem to have enjoyed a fairly normal life, roosting in the trees in the summer and sheltering in the stables and sheds during the winter with the other animals on the farm. Families used the birds for food and sold them and their eggs at the country store and to traveling haulers.

Live chicken haulers traditionally "went from farm to farm collecting cockerels [young roosters] and culls [sick, spent, and deformed birds] from the laying flock, establishing small feeding stations and assembling a sufficient quantity of birds to haul or ship to the big city markets. . . . The buyer, usually a live poultry broker, would take ownership after the birds were inspected and make arrangements for delivery to other live poultry handlers, city processors, or butcher shops." This is essentially the nature of the live poultry market trade up to this very day (Huntley).

A common practice was to fake the weight of the birds by such tricks as "feeding ingredients to bind the lower intestinal tract followed by feeding salt to encourage heavy water consumption. Poultry was also seeded with gravel or lead shot to increase weight, or fed heavily on a diet of corn just prior to unloading and weighing" (Watts and Kennett, 6).

Before the Second World War, women were the primary caretakers of poultry in the United States. According to *American Poultry History*, many men felt it was beneath them to "spend their time fussing with a lot of hens." Mrs. W. B. Morehouse told a Wisconsin Farmers' Institute audience in 1892, "A good many of the masculine gender tell us that it will so lower their dignity as to actually become a poultry keeper." On most farms, the housewife and children looked after the flock, using the pin money received to buy groceries. Early poultry extension programs were aimed at appealing to farm women. However, as poultry-keeping changed from a small farm project to a major business enterprise, it wasn't long until, as one woman put it, "my" flock became "our" flock and ultimately "his" flock (Skinner, 61, 75).

Until the 1920s, broody hens (true or foster mothers), and in some parts of the country surgically caponized (castrated) roosters, were used to rear young chickens in old-fashioned coops. During the 1920s, hatcheries with artificial incubators and brooders became widespread. Poultry husbandry classes and home economics curricula gave way to poultry science programs at land grant colleges and universities.

In the 1920s, feed companies like Ralston Purina, Quaker Oats, and Larrowe Milling, the forerunner of General Mills, set up poultry research facilities. The founding of Kimber Farms in 1934, in Fremont, California, launched the modern genetics research laboratory focusing on the breeding of chickens for specific economic traits such as heavy egg-laying. Kimber Farms developed a line of vaccines to cope with the chicken diseases that sprang in all directions as a result of genetic hybridization, which weakened disease resistance, and were exacerbated by the increased crowding and proximity of flocks to one another in chicken-producing areas (Smith and Daniel, 270–272).

Since the 1950s, chickens have been divided into two distinct genetic types—broiler chickens for meat production and laying hens for egg production. Battery cages for laying hens—identical units of confinement arranged in rows and tiers—and confinement sheds for broiler chickens came into standard commercial use during the 1940s and the 1950s. World War II, urbanization, and a growing human population produced a demand for cheap, mass-produced poultry and eggs. Following World War II, many dairy barns were remodeled into caged laying-hen facilities to meet the demand for poultry and eggs that grew during the war when these items were not rationed as red meat was (Skinner, 227).

By 1950, most cities and many villages had zoning laws restricting or banning the keeping of poultry, a pattern that facilitated the decline of breeding "fancy" fowl for exhibition in favor of breeding "utility" fowl for commercial food production. Poultry diseases proliferated with the growing concentration of the confined utility flocks that kept getting larger. As a result, traditional poultry-keeping and poultry shows both came to be viewed as potential disease routes, similar to current claims that chickens kept outdoors spread avian influenza viruses (GRAIN, 3–6). Then, as now, largely under the direction of the U.S. Department of Agriculture and its industrialized counterparts around the world, an increasingly intricate system of voluntary "sanitation," medication, and mass-extermination procedures was established to protect the poultry industry from the problems that the industry itself created.

Following World War II, the system known as vertical integration replaced earlier methods of chicken production. Under this system, a single company or producer, such as Tyson or Perdue, owns all production sectors, including the birds, hatcheries, feed mills, transportation services, medications, slaughter plants, further processing facilities, and delivery to buyers. The producer contracts with small farmers, known as "growers," of whom there are approximately 25,000 in the United States (Bell and Weaver, 814). Growers supply the land, housing, and equipment, look after the chickens, and dispose of the waste: the dead chickens and manure-soaked wood shavings and sawdust bedding on which the birds are raised, known as litter. In this way, a major capital investment, together with the burden of land and water pollution, is shifted to people whom the company can terminate practically at will, and who are often left with mortgages to pay off, scant savings, and little or no legal protection.

Despite the contaminated wells and inequities of this system, growers have traditionally been reluctant to complain for fear the company would stop sending them chickens. In 1992, poultry growers in the United States formed the National Contract Poultry Growers Association. Based in Sanford, North Carolina, the NCPGA campaigns for better treatment of growers through legislation and public education. On August 1, 2007, the USDA's Grain Inspection, Packers, and Stockyard Administration announced proposals that, if adopted, will make U.S. poultry growers more equal partners with the integrator companies they work for (Schuff 2007b).

Historically, the broiler chicken industry began in New England, then shifted to the South, where, in addition to the warm weather, there is little or no union activity, a large, undereducated rural population, and a receptive political climate that includes a "light tax and regulatory burden" (Bell and Weaver, 13). While some of the region's original attractions have succumbed to higher land and labor costs, the South's business climate, together with its "human capital and institutional infrastructure," continues to make it preferable to other parts of the country (Bell and Weaver, 809–814).

Along with better financial security, poultry growers, slaughterhouse workers, chicken catchers, "live-haul" drivers, and forklift operators desire a sense of dignity from the companies they work for. They resent being lumped together with the chickens. However, their wish runs counter to the history of the industry, which prides itself on having overcome the general attitude of appreciating individual male and female birds as well as individual farmers. The birds and the workers are "part of an efficient system of food production" (Skinner, 245).

Even today, at a time when U.S. poultry workers are starting to gain some attention, they are still, according to Steve Striffler in his book *Chicken*, "oddly incidental" to the industry that employs them (Striffler, 71). A sense of how many growers view themselves in relation to the companies they work for can be found in Sylvia Tomlinson's fictionalized account *Plucked and Burned*, where a grower complains to his neighbors:

> They've got a million frickin' ways to pluck us. Pardon my language, ladies, but I just got delivered 22,000 sorry biddies [chicks] for my houses that are supposed to hold 27,000. On top of that, besides being puny, I think most of them are blind. And it's certainly not from ammonia in the litter, which is what you'll hear the company claiming a couple of days from now. We did a cleanout down to the floor this time and put all new shavings in so we'd be in good shape before going into the winter. Those birds arrived half dead. If that's not retaliation for us trying to organize, then I don't know the meaning of the word" (Tomlinson, 180–181).

Behavior and Treatment of Chickens

The treatment of chickens for food in modern society is astonishingly ugly and cruel. The mechanized environment, mutilations, starvation procedures, and methodologies of mass-murdering birds, euphemistically referred to as "food production," raise many profound and unsettling questions about our society and our species. A former pharmaceutical company employee with the poultry industry wrote, "One of my worst experiences, and it didn't even involve live animals, was the World Poultry Expo in Atlanta. It horrified me because its energy and unquestioned acceptance paralleled a Holocaust concentration camp. It was upsetting to see how entrenched economically some very appalling practices are. I would walk through the aisles and think, 'I am probably one of the few people here (out of thousands) who find this disturbing'—and I found that very disturbing" (Farb 1993b).

Thus far, our responsibility for how badly we treat chickens and allow them to be treated has been dismissed with blistering rhetoric designed to silence objection: "How the hell can you compare the feelings of a hen with those of a human being?" However, animal scientists have begun challenging society's ignorance with studies showing the intelligence and sensitivities of chickens (Rogers 1995, 1997; Davis 2007).

Still, people are intimidated. We are told that we humans are capable of knowing just about anything we want to know—except, ironically, what it feels like to be one of our victims. We are told we are being "emotional" or "anthropomorphic" if we care about a chicken and grieve over a chicken's plight. However, it is not "emotion" as such that is under attack but the vicarious emotions of pity, sympathy, compassion, sorrow, and indignity on behalf of the victim, a fellow creature—emotions that undermine business as usual. By contrast, such "manly" emotions as patriotism, pride, conquest, control, and mastery are encouraged.

One of the main arguments that is used to silence opposition to the cruelty imposed on chickens on factory farms is that they are "productive"—only "happy" chickens lay lots of eggs and put on a ton of weight. In fact, chickens do not gain weight and lay eggs in inimical surroundings because they are comfortable, content, or well-cared for, but because they are specifically manipulated to do these things through genetics and management techniques that have nothing to do with happiness, except to destroy it. In addition, chickens in production agriculture are slaughtered at extremely young ages, before diseases and death have decimated the flocks as they would otherwise do, even with all the medications.

Notwithstanding, millions of young chickens die each year before going to slaughter, and on the way to slaughter, but because the volume of birds is so big—in the billions—the losses are economically negligible. Many more birds suffer and die on factory farms than on traditional farms; however, more pounds of flesh and eggs are realized under it also. The term "productivity" is an economic measure referring to averages, not the well-being of individuals (Mench 1992, 108). Excess fertility and musculature are not the criteria that we use to judge the well-being of human beings, and they are not indices of avian well-being either. They more likely signify the opposite.

Chickens are not suited to the captivity imposed on them to satisfy human wants in the modern world. Michael W. Fox, a veterinary specialist in farmed-animal welfare, states that while chickens and other factory-farmed animals may sometimes appear to be adapted to the adverse conditions under which they are kept, "on the basis of their functional and structural 'breakdown,' which is expressed in the form of various production diseases, they are clearly not adapted" (Fox, 209).

Marxian Alienation of Factory-Farmed Chickens

Barbara Noske, in her book *Humans and Other Animals*, notes that there is no compelling reason why nonhuman animals should not be regarded by humans as "total beings whose relations with their physical and social environment are of vital importance" (1989, 18). The morality of forcing human beings to subsist in an alien environment to serve economic objectives was analyzed by Karl Marx in terms that provide insight into the experience of chickens shunted into human-created environments that are alien to their nature. Marx described four interrelated aspects of alienation: from the product, from the productive activity, from the species life, and from nature. We can look at chickens and other captive animals from a similar viewpoint.

From Their Own Products

Factory-farmed chickens are alienated from their own products, which consist of their eggs, their chicks, and parts of their own bodies. The eggs of chickens used for breeding are taken away to be artificially incubated and hatched in mechanized hatcheries, and those of caged laying hens roll onto a conveyer belt out of sight. Parents and progeny are severed from one another. Factory-farmed chickens live and die without ever knowing a mother. The relationship between the chicken and his or her own body is perverted and degraded by factory farming. An example is the cruel conflict in young broiler chickens between their abnormally rapid accumulation of breast muscle tissue and a developing young skeleton that cannot cope with the weight, resulting in crippling, painful hip joint degeneration and other afflictions that prevent the bird from walking normally, and often, or finally, from walking at all. Human sufferers can obtain pain relief medication; the chickens have no such options.

From Their Own Productive Activity

Chickens are alienated from their own productive activity, which is reduced to the single biological function of laying eggs or gaining weight at the expense of the whole bird. Normal species activity is prevented so that food will be converted into this particular function only and not be "wasted" (Bell and Weaver, 866). Exercise of the chicken's natural repertoire of interests and behaviors conflicts fundamentally with the goals of factory farming.

From Their Own Societies

Chickens are alienated from their own societies. Their species life is distorted by crowding and caging, by separation of parents and offspring, by the huge numbers of birds crowded into a vast confinement area—somewhat as if one were compelled to live one's entire life at an indoor rock concert or political rally, after the show was over—and by the lack of natural contact with other age groups and sexes within the species. Chickens should be living in small groups that spend their days foraging for food, socializing, and being active; thus, the egg industry will cynically tell you that one of the advantages of the battery cage is that it satisfies the chicken's need to be part of a little flock. More aptly, Michael Baxter writes, "Forcing hens into such close proximity as occurs in a battery cage disrupts normal social interaction and suggests that the hens continually strive to get further apart. The ongoing regulation of social spacing and continuous awareness of other hens in the cage provide evidence of social conflict and indicate that hens are stressed by being housed so close together" (618).

From the Natural World

In the most encompassing sense, factory-farmed chickens are alienated from surrounding nature, from an external world that answers intelligibly to their inner world. There is nothing for them to do or see or look forward to; no voluntary actions are permitted, or joy, or zest of living. They just have to *be*, in an existential void, until we kill them. The deterioration of mental and physical alertness fostered by these conditions has been suggested by some animal scientists as an adaptive mechanism offsetting the occurrence of long-term suffering. F. Wemelsfelder explains more reasonably: "It would be conceptually meaningless to assume that such states could in any way come to be experienced by an animal as 'normal' or 'adapted.' Behavioural flexibility represents the very capacity to achieve well-being or adaptation; impairment of such capacity presumably leaves an animal in a helpless state of continuous suffering" (122).

Chicken specialist Lesley Rogers states that while an ultimate aim of captive breeding programs is to obtain birds with minds "so blunted that they will passively accept overcrowded housing conditions and having virtually nothing to do but eat—and then to eat standard and boring food delivered automatically," there is "no evidence that domestic chickens, or other domestic breeds, have been so cognitively blunted that they need or want no more behavioural stimulation than they receive in battery farms" (1997, 184–185).

Of the suffering of hens in battery cages, Rogers writes that they not only suffer from restricted movement, they have "no opportunity to search for food and, if they are fed on powdered food [which they are], they have no opportunity to decide which grains to peck. These are just some examples of the impoverishment of their environment. . . . Chickens experiencing such environmental conditions attempt to find ways to cope with them. Their behavioural repertoire becomes directed towards self or cage mates and takes on abnormal patterns, such as feather pecking or other stereotyped behaviors . . . used as indicators of stress in caged animals" (1995, 219).

I've seen signs of this kind of stress in our household chickens. In addition to their other expressive languages, chickens have a piping voice of woe and dreariness whenever they are bored or at a loose end. Occasionally, one of our hens has to be kept indoors to recover from an illness or because she is a new hen who has not yet joined the flock outside. Wearily, she will wander about the rooms, fretting and sometimes biting at my ankles, or tag disconsolately and beseechingly behind me, yawning and moaning like a soul in the last stages of ennui.

Reactions to the "Animal Machine"

At the automated hatchery the "contents of the trays (chicks, shells, eggs) are tipped onto the conveyer or onto a belt conveyer which then transports the chicks, eggs, shells, etc. to the separating roller conveyer. As the birds spread out across these oscillating rollers, they fall through the opening between the rollers to a belt or rod conveyor, which removes them from the machine. This process separates the chicks from the unhatched whole eggs and larger eggshells, as they are transported to the waste removal system."

—Bell and Weaver, *Commercial Chicken Meat and Egg Production*, 698–699

Some critics have argued that the revulsion we feel at how chickens and other animals are treated for food is not necessarily moral but perhaps only aesthetic. The "animal machine" offends our aesthetic consciousness. J. Baird Callicott writes: "The very presence of animals, so emblematic of delicate, complex organic tissue, surrounded by machines, connected to machines, penetrated by machines in research laboratories or crowded together in space-age 'production facilities' is surely the more real and visceral source of our outrage at vivisection and factory farming than the contemplation of the quantity of pain these unfortunate beings experience" (335).

In this view, we do not identify with the animals or their pain, or burden our thoughts with the misery of their lives at our hands. Rather, our reactions are produced by something more abstractly incongruous of which the situation including the animal is "emblematic." Robert Burruss writes somewhat more searchingly:

> About 20 years ago, *Scientific American* ran an article on the management of chickens in the production of eggs and meat. Concentration camps for chickens is what one friend who read the article called the chicken farms.
>
> My enjoyment of eggs and chicken has forever been abridged by that article. . . . [T]he problem is not moral; rather it is . . . the images evoked by the idea of scrambled eggs or chicken meat, images from the article of the ways the animals spend their bleak lives.
>
> Maybe, thinking about it now as I write this, those images actually are a basis of a moral judgment. Maybe that's how moral judgment originates.

Maybe.

A few years ago, a friend of mine was driving one afternoon down a back road on the Eastern Shore of Maryland when she came upon a chicken house, which she described as "in the middle of nowhere." She stopped the car, got out, walked over, unlatched the door, and tiptoed inside. There was the usual scene, thousands of young chickens, amid the ammonia haze, with the mechanical feeders and drinkers. Over in a corner, she noticed that some kind of exciting activity was taking place, and making her way over carefully, she saw that the birds in the immediate vicinity had either found, or else they had made, a hole in the ground through which they were crawling in and out to dustbathe.

Outside, around back, she watched the scene. She watched the young chickens as they threw up their little clouds of dust against the big sky, and the flat fields, and the long, low building with a sign that said, simply, "There is no one here but us chickens."

No. There was a witness. And through her eyes, I too became a witness to their lives.

CHAPTER 2

THE BIRTH AND FAMILY LIFE OF CHICKENS

Then they all settled down in the soft green shade of the lemon tree, with each little chick taking its turn to fly up to the best and softest seat on Granny Black's back. And while they waited for the sun to go down again, she told them about the great big world outside the chick run, or the days when she was a chick, or the story they liked telling best of all—her Miracle story about Eggs. How the broken fragments they had hatched from were once smooth, complete shapes; how every chicken that ever was had hatched out in exactly the same way; how only chooks could lay such beauties; and how every time they did, they were so filled with joy that they could not stay quiet, but had to burst into song; and how their song was taken up by England the cock and echoed by every single hen in the run.

—Mary Gage, *Praise the Egg* (11)

WHEN LIVING CREATURES BECOME "UNITS"

The birth of a chicken is a poignant event. In *The Chicken Book*, Page Smith and Charles Daniel write: "As each chick emerges from its shell in the dark cave of feathers underneath its mother, it lies for a time like any newborn creature, exhausted, naked, and extremely vulnerable. And as the mother may be taken as the epitome of motherhood, so the newborn chick may be taken as an archetypal representative of babies of all species, human and animal alike, just brought into the world" (321).

Most of us know deep inside that we are members of a single family of living creatures, yet many people resist this knowledge and its implications. Evolution is accepted, but the sentiment of kinship still struggles to evolve. Once, I was reproved by a former meat inspector for issuing a news release that in his view ignored "hard science" and sentimentalized chickens in order to win sympathy for their plight (Leonard). I had stated, "For a chicken trapped inside the world of modern food manufacturing, to break out of the shell is to enter a deeper darkness full of bewildering pain and suffering from birth to death" (UPC 1991). I noted that a mother hen will tenderly and even fiercely protect her young brood, driving off predators and sheltering her little chicks beneath her wings, and that roosters will often join in the hen's egg-laying ritual, which is an extremely important and private part of a chicken's life.

While dismissing these statements as "unscientific," the inspector acknowledged the justness of my and others' descriptions of the "visceral horrors of an ordinary day at the slaughterhouse, where humans and birds are often treated in inhumane ways." Especially disturbing to him was the egg industry's treatment of male chicks, who on hatching are thrown into trashcans to suffocate.

Clearly a struggle was taking place here between recognition of the link between chickens and humans—which alone would explain why both groups could be judged as inhumanely treated—and the dogma that chickens (and virtually all other nonhuman creatures) do not have experiences comparable to human experience. Manifest similarities between their behavior and ours, as in the parental care and protection of offspring, are dismissed as "mere instinct" in them, even though human behavior is similarly grounded in the instinctual impulses and corresponding patterns of emotion that characterize our own species and link our species to others.

Observation of natural incubation has shown that a hen turns each egg as often as 30 times a day, using her body, her feet, and her beak to move her eggs in order to maintain the proper temperature, moisture, ventilation, humidity, and position of the egg during the 21-day incubation period (Skinner, 140). Though new to the West, artificial incubation of eggs has been practiced for over 2,000 years in Egypt, China, and other Eastern countries. However, none of these procedures "which involves the use of artificial incubation, seems to mimic closely natural incubation by the hen" (Rogers 1995, 70).

The automated poultry-slaughtering technology developed in the

1940s and 1950s followed the development of mechanical incubators at the turn of the century. Mechanical incubators, in hatcheries that are now capable of hatching 1,400,000 chicks per week out of 1,650,000 eggs set in incubator trays (Bell and Weaver, 700, 822), enabled a farmer to start with 100 or more baby chicks without requiring a hen to sit on a nest and hatch her chicks. The development of huge hatcheries dispensed with the hen's warmth and nurturing as well. Henceforth, the hen would be a "breeder" or a "layer" instead of a mother.

School Hatching Projects

My kids are incubating eggs. One hatched this morning about 12 hours ago. It is still partially wet and breathing hard and not doing a lot of walking. What can we do to help it along or is this somewhat routine? Also, in breaking out of his shell he upset the other eggs that were in individual egg compartments, which we cut out from an egg carton. Should we put the eggs back into their compartments or leave them where they ended up?

—P. Maloney, teacher, in an e-mail to
United Poultry Concerns, April 19, 2005

Ethical questions are raised when unwanted animals are brought into this world, diminishing our sense of the inherent value of the living creature. The positive lesson that can come from observing and respecting normal parenting of adult birds for their future offspring is lost.

—F. Barbara Orlans, Senior Research Fellow at the Kennedy
Institute of Ethics, Georgetown University, Washington, DC

Not surprisingly, many people in the modern world fail to perceive chickens as even having a mother, let alone a father. The school hatching programs that began in the 1950s mislead children and teachers to think that chickens come from mechanical incubators. Supplemental facts about the role of the rooster and the hen, even if provided, cannot compete with the mechanized classroom experience. Each year, kindergarten and elementary school teachers and their students place thousands of fertilized eggs in classroom incubators to be hatched within three to four weeks. They're encouraged to do this by the school district's science coordinators and the biological supply companies, which advertise fertile eggs and "easy-to-use" incubators in the catalogs they send to the schools and display at teachers' conventions. In 1994, one egg supplier sold 1,800 eggs to New York City schools alone (Miller 1994).

More than a decade later, school hatching projects continue in urban areas like Montgomery County, Maryland, a suburb of Washington, DC, and New York City, where keeping chickens is either illegal or impossible, and animal shelters and sanctuaries have to scramble to accommodate the forlorn victims of "miracle of birth" projects each year. As a humane society spokesperson told the *Potomac Gazette* about the sad survivors who are routinely dumped at the shelter by the schools, "I'm not seeing people who are looking for them" (Rathner, A-11).

These birds are not only deprived of a mother; many grow sick and deformed because their exacting needs are not met during incubation and after hatching. Chick organs stick to the sides of the shells because they are not rotated properly. Chicks are born with their intestines outside their bodies. Eggs hatch on weekends when no one is in school to care for the chicks. The heat may be turned off for the weekend, causing the chicks to become crippled or die in the shell. Some teachers even remove an egg from the incubator every other day and open it up to look at the chick in various stages of development, adding the killing of innocent life to the child's education (Ubinas).

Biologist F. Barbara Orlans writes in *Hatching Good Lessons*: "In these projects, any sense of parent birds carefully preparing nests and tending their future babies is lost because the eggs are hatched in a piece of equipment. The surviving chicks are usually doomed to a life expectancy of a few days spent miserably. Young birds need nurturing and rest. They are difficult to feed in the classroom and can suffer starvation and dehydration that is not even noticed."

When the project is over, the survivors must be disposed of. Because a child bonds naturally with young animals, students are told, and some teachers want to believe, that the chicks are going to live "happily on a farm." In reality, most of them are going to be killed immediately (working farms do not assimilate school-project birds into their flocks for fear of disease), sold to live poultry markets and auctions, fed to captive zoo animals, or left to die slowly of hunger and thirst as a result of ignorance and neglect. Some of the birds end up in shelters where they are either destroyed or, in rare cases, adopted out.

Urbanization and zoning laws enormously compound the problem. Most residential zoning laws ban the keeping of domestic fowl, and even people who can provide a good home for a chicken can accommodate only so many roosters. Normal flocks have several female birds to one male, roosters crow before dawn and periodically during the day, and some roosters will attack people to protect their hens. Unfortunately, due to "unequal mortality of the sexes during embryonic development," more than half of all chickens born are males (Bell and Weaver, 737).

Chick hatching projects teach children and teachers that bringing a life into the world is not a grave responsibility with ultimate consequences for the life created. Children's public television has contributed to this desensitization and to the fallacy that chickens have no natural origin or need for a family life. *The Reading Rainbow* public television program "Chickens Aren't the Only Ones," based on a book by Ruth Heller, shows that other kinds of animals besides chickens lay eggs. However, chickens are the only ones represented in barren surroundings. One heartless scene shows a baby chick struggling out of its egg alone on a bare table, while ugly, insensitive music blares, "I'm breaking out."

The *3-2-1 Contact* show "Pignews: Chickens and Pigs" has aired frequently on children's public television. Promoting the agribusiness theme of "changing nature to get the food we eat," it shows hatchery footage of newborn chicks being hurled down stainless steel conveyers, tumbling in revolving sexing carousels, flung down dark holes, and brutally handled by chicken sexers who grab them, toss them, and hold them by one wing while asserting that none of this hurts them at all. These scenes alternate with rapid sequence images of mass-produced fruits and vegetables. Children are brightly told that "farmers are changing how we grow 100 million baby chicks a week, 3 million pounds of tomatoes, 36 billion pounds of potatoes." Chickens are described against a background of upbeat music as a "monocrop" suited to the "conveyer belt and assembly line, as in a factory."

Is it any wonder that many people regard chickens as some sort of weird chimerical concoction comprising a vegetable and a machine?

The Egg and Chick: Historical Symbols of Nature and Rebirth

Notwithstanding the seventeenth-century French philosopher and mathematician René Descartes' claim that nonhuman animals are machines, throughout history, the chick and the egg have symbolized the mystery of birth and renewal of life. The Italian Renaissance ornithologist Ulisse Aldrovandi wrote of the use of eggs in religious ceremonies: "Eggs were believed to reproduce all nature and to have a greater power for placation in religion and for prevailing upon the powers of heaven" (Smith and Daniel, 166). Hindus saw the beginning of the world as an enormous cosmic egg that incubated for a year and then split open, half silver and half gold. "The silver half became the earth; the gold, the sky; the outer membrane, mountains; the inner, mist and clouds; the veins were rivers, and the fluid part of the egg was the ocean, and from all of these came in turn the sun" (Smith and Daniel, 184).

The egg has been a traditional feature in many ancient rites of spring. Christianity adopted the egg as a symbol of Christian rebirth. The eggshell symbolized the tomb from which Christ had risen, and the inner content of the egg symbolized the theme of resurrection and hope for eternal life. According to *The Chicken Book*, the word "Easter" comes from "the name of the Anglo-Saxon goddess of spring, Eoestre, whose festival was on the first Sunday after the full moon following the vernal equinox." Eoestre is depicted in an ancient Anglo-Saxon statue holding an egg, the symbol of life, in her hand (Smith and Daniel, 184–185, 272).

Easter Egg Hunt and Egg Gathering

The association of the hen's egg with Easter and spring survives ironically in the annual children's Easter egg hunt, for the origin of this ritual has been largely forgotten. Traditionally, the finding of eggs was identified with the finding of riches. The search for eggs was a normal part of farm life, because a free hen sensibly lays her eggs in a sheltered and secluded spot. However, today's children hunt for eggs that were laid by a hen imprisoned in a wire cage in a mechanized building. The widespread disappearance of the home chicken flock in the 1950s ended the gathering of eggs laid by a hen in the place that she chose for her nest. Historian Page Smith writes in *The Chicken Book*, "My contemporaries who have such dismal memories of chickens from the unpleasant chores of their youth had experienced already the consequences of putting living creatures in circumstances that are inherently uncongenial to them" (Smith and Daniel, 272).

Wilbor Wilson provides the background to this change in *American Poultry History*. He writes: "As the size of poultry ranches increased, the chore of egg gathering became drudgery instead of pleasure. Rollaway nests with sloping floors made of hardware cloth offered a partial solution, but the number of floor eggs increased when the hens did not readily adopt the wire-floored nests. This changed with development of the cage system, which incorporated the roll-out feature and left the hen no choice" (Skinner, 234–235).

The Hen as a Symbol of Motherhood

[T]he continued emphasis genetically [has been] on smaller, more efficient but lighter-weight egg machines.

—Skinner, *American Poultry History*, 367

In our day, the hen has been degraded to an "egg machine." In previous eras, she embodied the essence of motherhood. The first-century AD Roman historian and biographer Plutarch wrote praisingly of the mother hen in *De amore parentis* [*parental love*]: "What of the hens whom we observe each day at home, with what care and assiduity they govern and guard their chicks? Some let down their wings for the chicks to come under; others arch their backs for them to climb upon; there is no part of their bodies with which they do not wish to cherish their chicks if they can, nor do they do this without a joy and alacrity which they seem to exhibit by the sound of their voices" (Smith and Daniel, 27).

In Matthew 23:37, the mother hen is evoked to express the spirit of yearning and protective love in Christ's lament: "Jerusalem, Jerusalem, how often have I wished to gather your children together, as a hen gathers her chicks, and you did not wish it."

The Renaissance writer Aldrovandi wrote of mother hens in the sixteenth century:

> "They follow their chicks with such great love that, if they see or spy at a distance any harmful animal, such as a kite or a weasel or someone even larger stalking their little ones, the hens first gather them under the shadow of their wings, and with this covering they put up such a very fierce defense—striking fear into their opponent in the midst of a frightful clamor, using both wings and beak—they would rather die for their chicks than seek safety in flight. . . . Thus they present a noble example in love of their offspring, as also when they feed them, offering the food they have collected and neglecting their own hunger" (Smith and Daniel, 162).

Maternal Instincts in the Domestic Hen

While the egg industry claims that the modern "egg machine" has had the broodiness bred out of her, it is more likely that the hen's mothering impulses have been suppressed rather than eliminated. Jennifer Raymond wrote of her surprise on purchasing a hen by mail order:

> Another benefit of the White Leghorn, according to the Sears catalogue, is that the maternal instinct has been bred out of the hens so they don't "go broody." Going broody is the notion hens get to sit on eggs and raise a family. During this time, hens stop laying. Needless to say, this tendency has no commercial value. One of my hens seemed to be a throwback, however, and began spending all her time in the hen house sitting on the nest.
>
> Since I had no rooster, the eggs weren't fertile, and her efforts would have proven futile had I not procured some fertile eggs from my neighbors and placed them in the nesting box. Nineteen days later, I woke to see her out in the yard, followed by five little red balls of fluff. She was an attentive mother, teaching the chicks to scratch and showing them all the best places to look for food. Soon the chicks were as large as their "mother," but they still gathered underneath her at night. It was so comical to see these large, gawky adolescent youngsters sticking out on all sides of the little white hen.

Scientists have documented the revival of maternal behavior in feral hens. (The term "feral" refers to domesticated animals who return to a self-sustaining way of life.) For example, in a classic field study of feral chickens on a coral island northeast of Queensland, Australia, in the 1960s, animal researcher G. McBride and his colleagues recorded the birds' social and parental behavior over the course of a year, including the behavior of a hen and her chicks in a moment of human disturbance:

> When a broody hen with very young chicks is disturbed by a man, the hen gives a full display and the alarm cackle. When pressed closely, the hen hides her chicks in the following way: she regularly turns and makes a short charge at her pursuer. As she turns, she pushes one or two of her chicks into a hollow, while giving a particularly loud squawk among the clucks. If the chick finds the hollow, it remains still while extremely well camouflaged. If not hidden, it gives a strong distress chirp and the dam [hen] returns for it (McBride, 140).

Like their ancestors and contemporary wild relatives in the tropical forests, feral chickens form "small, discrete social groups" that spend "much of their day foraging either separately or together, then returning at dusk to roost." Hens conceal their nests, raise and defend their broods. "In short," as Nicol and Dawkins wrote in *New Scientist*, "there is no evidence that genetic selection for egg laying has eliminated the birds' potential to perform a wide variety of behaviour" (1990, 46).

The Role of the Rooster

The family role of the rooster is even less well known to most people than the motherhood of the hen. The charm of seeing a rooster with his hens appears in Chaucer's portrait of Chanticleer in *The Canterbury Tales*:

> This cock had in his princely sway and measure
> Seven hens to satisfy his every pleasure,
> Who were his sisters and his sweethearts true,
> Each wonderfully like him in her hue,
> Of whom the fairest-feathered throat to see
> Was fair Dame Partlet. Courteous was she,
> Discreet, and always acted debonairly.

In ancient times, the rooster was esteemed for his sexual vigor; it is said that a healthy young cock may mate as often as 30 or more times a day (Bell and Weaver, 652). According to *The Chicken Book*, "The extreme erectness of the cock, straining upward, has suggested to many besides the Greeks the erectness of a tumid penis" (Smith and Daniel, 53). The rooster thus figures in religious history as a symbol of divine fertility and the life force. In his own world of chickendom, the rooster—the cock—is a father, a lover, a brother, a food-finder, a guardian, and a sentinel.

Aldrovandi extolled the rooster's domestic virtues:

> He is for us the example of the best and truest father of a family. For he not only presents himself as a vigilant guardian of his little ones, and in the morning, at the proper time, invites us to our daily labor; but he sallies forth as the first, not only with his crowing, by which he shows what must be done, but he sweeps everything, explores and spies out everything.
>
> Finding food, "he calls both hens and chicks together to eat it while he stands like a father and host at a banquet . . . inviting them to the feast, exercised by a single care, that they should have something to eat. Meanwhile he scurries about to find something nearby, and when he has found it, he calls his family again in a loud voice. They run to the spot. He stretches himself

> up, looks around for any danger that may be near, runs about the entire poultry yard, here and there plucking up a grain or two for himself without ceasing to invite the others to follow him" (Smith and Daniel, 65).

A nineteenth-century poultry keeper wrote to his friend that his Shanghai cock was "very attentive to his Hens, and exercises a most fatherly care over the Chicks in his yard. . . . He frequently would allow them to perch on his back, and in this manner carry them into the house, and then up the chicken ladder" (Smith and Daniel, 216).

In the McBride study cited above, the rooster with his hens and their young are described:

> When a group moved, it was the male who gathered the females together before moving. The hens maintained contact with him while moving, and he controlled the movement when it crossed open ground. When disturbed, he gave the alarm call and walked parallel to the predator or potential predator while the hens quietly hid. When the flock was disturbed, males were actually observed to drive the females away by rushing toward them with wings spread. While hens fed, males spent the majority of the time on guard in the tail-up, wing-down alert posture. . . . Males used the typical broody hen display when charging, tail fanned, wings down, and feathers puffed. Both went to roost in the trees at night and called the brood or flock to them (McBride, 143).

Why Roosters Crow

In all the realm of crowing he had no peer. . . .
Stouter was this cock in crowing than the loudest abbey clock.
Of astronomy instinctively aware,
He kept the sun's hours with celestial care.

—Chaucer on Chanticleer

The thing most people identify with roosters is crowing. Why do roosters crow? Remember that chickens are originally from the jungle. Their wild relatives have lived in tropical forests for tens of thousands of years. Perched in the trees, and sensitive to infrared light, chickens see morning light at least 45 minutes before we do. (Sullenberger 1994, 2). Indeed, as Lesley Rogers writes: "Chickens have well-developed vision with spectral sensitivities ranging from the infrared to the ultraviolet regions of the spectrum, and much of this visual capacity is known to develop before hatching (Rogers 1995, 27).

Chickens also have very keen ears, a distinct advantage when living in deep foliage, where it can be difficult to see predators and keep track of one's flock when the sub-flocks are constantly foraging from place to place. According to chicken expert David Sullenberger, "Sight is of limited value in the dense cover of a jungle, especially in regard to knowing where one is in relation to the other members of the flock. Sound, however, works very well in dense cover. Enter the crow!" (Sullenberger 1993, 6).

Through their crowing, every rooster knows where every other rooster is at all times. Each rooster can recognize the crow of at least 30 other roosters, maybe more (Sullenberger 1993, 6). As protectors of the flock, roosters are always on the lookout. If a rooster spots danger, he sends up a shrill cry. The other roosters echo the cry. Thereupon, the whole flock will often start up a loud, incessant, drum-beating chorus with all members facing the direction of the first alarm, or scattering for cover in the opposite direction.

When it looks safe again, an "all clear?" query goes out from the rooster, first one, followed by the others, in their various hiding places. Eventually, the "all clear" crow is sent up by the bird who first raised the alarm, and a series of locator crows confirms where every other rooster and his sub-flocks are at that point.

Relationship Between the Rooster and the Hen

Mating and nesting elicit other kinds of vocal communication within the flock. When a hen is ready to lay an egg, she gives a nesting call, inviting her mate to join her in finding a nest site. Together, the hen and rooster find and create a nest by pulling and flinging around themselves twigs, feathers, hay, leaves, and loose dirt, after they have scraped a depression with their beaks and feet. But first comes the search.

When the rooster finds a place he likes, under a log, perhaps, he settles into it and rocks from side to side while turning in a slow circle and uttering primeval grumbling growls, which may or may not convince the hen that this is the place. She may accept it, or they may look for another site. During and after the search, the hen cackles and squawks to keep the rooster coming back to her while she is away from the protection of the flock (McBride, 165–168, 179).

Often I have heard one of our hens call out to her rooster partner: "I'm all alone. Get over here!" Our normally quiet hen, Petal, raised a ruckus if her adored Jules was out of her sight for long, even if she had not just laid an egg. Her otherwise demure little voice became SQUAWK, SQUAWK, SQUAWK. Jules would lift his head up, straighten up, mutter to himself in what can only be described as "chicken talk," and do an about-face. Off he went to comfort Petal. Silence.

Each of our "broiler" roosters, Henry and Phoenix, stood by his favorite hen while she laid her egg, and nowadays I watch Sir Valery and Curtis do likewise. I've even seen the whole little flock gather around a nesting hen in one of our chicken houses for half an hour or longer until her egg was laid. Once when I was in the car with Phoenix, a man ran over to us in the parking lot and said, "When I was a young man I worked on a chicken farm, and do you know, one of the most amazing things about those chickens was that they would actually choose each other and refuse to mate with anyone else."

Though chickens are polygamous, mating with more than one member of the opposite sex, individual birds are attracted to each other (Bell and Weaver, 652). They not only "breed"; they form bonds, "always sharing their goodies and clucking endearments to one another throughout the day." A rooster does a courtly dance for his special hens in which he "skitters sideways and opens his wing feathers downward like Japanese fans—the chicken version of the strut that is found in many bird species" (Luttmann, 46).

Bravery of Chickens

The call of the wild is in the chicken's heart, too. Far from being "chicken," roosters and hens are legendary for bravery. In classical times, the bearing of the cock symbolized military valor: his crest stood for the soldier's helmet and his spurs stood for the sword (Smith and Daniel, 66). A chicken will stand up to an adult human being. Our tiny Bantam rooster, Bantu, would flash out of the bushes and repeatedly attack our legs, his body tense, his eyes riveted on our shins, lest we should threaten his beloved hens.

An annoyed hen will confront a pesky young rooster with her hackles raised and run him off! Although chickens will fight fiercely and successfully with foxes and eagles to protect their family, with humans such bravery usually does not win. A woman employed on a breeder farm in Maryland wrote a letter to the newspaper berating the defenders of chickens for trying to make her lose her job, threatening her ability to support herself and her daughter (Sadler).

For her, "breeder" hens were "mean" birds who "peck your arm when you are trying to collect the eggs." In her defense of her life and her daughter's life against the champions of chickens, she failed to see the comparison between her motherly protection of her child and the exploited hen's courageous effort to protect her own offspring.

In an outdoor chicken flock, ritual and playful sparring and chasing normally suffice to maintain peace and resolve disputes without actual bloodshed. Even hens occasionally have a go at each other, but in 20 years of keeping chickens, I have never seen a hen-fight, with its ritualized postures and gestures, turn seriously violent or last for more than a few minutes. Chickens have a natural sense of order and learn quickly from each other. An exasperated bird will either move away from the offender or aim a peck, or a pecking gesture, which sends the message "lay off!" or "back off!"

Bloody battles, as when a new bird is introduced into an established flock, are rare, short-lived, and usually affect the comb (the crest on top of a chicken's head), which, being packed with blood vessels, can make an injury look worse than it is. It's when chickens are crowded, confined, bored, or forced to compete at a feeder that distempered behavior can erupt. However, chickens allowed to grow up in successive generations unconfined do not evince a rigid "pecking order" (Smith and Daniel, 165–166, 316). Parents oversee their young, and the young contend playfully, among many other activities. A flock of well-acquainted adults is an amiable social group.

Sometimes chickens run away; however, fleeing from a bully or hereditary predator-species on legs designed for the purpose does not constitute cowardice.

Formation and Laying of the Egg

A nesting hen is a comforting sight, as shown by the frequency of this image in decorative art. However, the setting hen is no more idle than the lilies of the field that, while appearing "neither to toil nor spin," are invisibly active. The hen turns her eggs many times a day and keeps her nest fresh and clean. If an egg rolls away, she pulls it back under her with her beak. In addition, she leaves the nest for 10 to 20 minutes each day to forage for food, drink water, defecate, and stretch her wings. Artificially incubated eggs must be cooled for 15 to 20 minutes a day to match the time the hen is away from her nest (Smith and Daniel, 320).

A rooster is not required for a hen to lay eggs. Eggs are periodically shed from her body the same as in other vertebrate females. However, the avian female has but one mature ovary, the left, and it is large in relation to the rest of her body compared to the ovaries of a mammal. In addition, a hen's ovum is surrounded by the yolk, albumin, shell membranes, shell, and cuticle necessary to nourish and protect the life of an embryo developing outside the mother. An egg is the female component of the species germline and is therefore present in some form at all stages. As noted in *The Chicken Book*, "Even when the chick is in the egg there are eggs within the chick, microscopically small but full of potential" (Smith and Daniel, 170).

Of the thousands of ova, only a small number actually mature to be laid, fertile or otherwise. A hen lays a group of eggs, one egg a day, in an indeterminate sequence of three to fifteen eggs at a time, called a clutch. The eggs of the sequence are often laid a little later each day, starting in the early morning an hour or two after sunrise; thus an egg laid late in the afternoon signals the end of a sequence. Then the hen skips a day or more of ovulation and egg laying before starting another clutch. If the eggs are fertile (the result of the hen's mating with a rooster), she waits to incubate them until the last egg of the clutch is laid, thus ensuring that all the eggs start to develop and hatch at the same time (Burton, 10).

Like the egg of a mammal, a hen's egg consists of a tiny reproductive cell, called a blastoderm, from which the embryo develops. In the chicken, the blastoderm is surrounded by the embryo's food, or yolk, and subsequent overlays. It takes about 10 days for the yolk to mature. This is followed by ovulation, at which time the mature yolk bursts from the ovary to be seized and engulfed by the funnel-shaped opening to the oviduct, called the infundibulum, which partially surrounds the ovary. Fleshly projections from the oviduct fill with blood, and the walls of the oviduct writhe and contract, moving the rotating egg into the portion of the oviduct known as the magnum. Here it receives the white, or albumin, the first layer of which becomes twisted at each end in opposite directions. These twisted ends, called the chalazae, polarize the egg and centralize the yolk after the egg is laid.

After two to three hours in the magnum, the egg goes to the isthmus to acquire the thin inner and thick outer shell membranes, composed of tough protein fibers, which prevent bacteria and other organisms from entering the egg. These membranes are in contact everywhere except at the large end of the egg, at the point where the air cell appears soon after the egg is laid.

After about an hour in the isthmus, the egg travels to the shell gland, or uterus, where it remains for 18 to 20 hours. Here, water and salts penetrate the shell membranes by osmosis, and the egg is kneaded by the muscular rhythms of the uterus into its final shape as the calcium salts are deposited. There are two layers: an inner shell composed of sponge-like calcite crystals, and an outer shell composed of hard, chalky calcite crystals about twice as thick as the inner shell crystals. The outer shell contains the brown, blue, green, or speckled color. Otherwise the shell is white. Color is based on molecular pigments, called the porphyrins, produced in the uterus when the shell is produced.

It takes between 23 and 26 hours for the egg to traverse the oviduct, including the vagina, where the cuticle is deposited, to be laid. If no sperm are present, either in the infundibulum or in the short, tubular projections in the lower portions of the oviduct, the egg will be infertile. Once the outer layers of white and shell surround the yolk, sperm cannot enter the ovum, though it may be stored in the hen for up to four weeks for fertilization (North and Bell, 46).

The actual laying of the egg is a complex process involving nervous signals from the brain to the muscles of the uterus and vagina, and the influence of hormones released from the posterior pituitary gland. Just as prolactin and other hormones that initiate maternal behavior are the same in both mammals and birds, so the hormones that stimulate muscular contractions in birds are the same ones that stimulate the uterine contractions in mammals leading to birth. This commonality, as noted in *The Chicken Book*, is one of many biological signals showing that despite evolutionary divergences, "chickens, and ourselves, are still members of a family, and a single family at that, of living creatures" (Smith and Daniel, 183–184).

Normally, the egg is in the hen's vagina for a few minutes, though it may reside there for several hours if necessary. The egg moves through the oviduct small end first, but just before oviposition it rotates horizontally in order to be laid with the large end first. This enables the uterine muscles to exert greater pressure on more surface area as the egg is being expelled. Finally, "in what is so obviously for the hen a moment full of pride and satisfaction, the egg, magnificently completed, is laid" (Smith and Daniel, 180).

If pride and satisfaction are an important part of egg laying in chickens, then the following description of the caged hen's ordeal may be cited in contrast:

> The frightened battery hen starts to panic as she vainly searches for privacy and a suitable nesting place in the crowded but bare wire cage; then she appears to become oblivious to her surroundings, struggling against the cage as though trying to escape. . . .
>
> Take a moment to imagine yourself as a layer chicken; your

> home is a crowded cage with a wire floor that causes your feet to hurt and become deformed; there's no room to stretch your legs or flap your wings and they become weak from lack of exercise; but at the same time, you can never be still because there is always one of your miserable cell mates who needs to move about; one of the other chickens is always picking on you and you cannot get away—except by letting others sit on top of you; the air is filled with dust and flying feathers that stick to the sides of the cage splattered with chicken shit from the inmates in the cage upstairs; it is hard to breathe—there is the choking stench of ammonia in the air from the piles of manure under the cages and you don't feel at all well; the flies are unbearable. . . . [E]ventually, despite your wretchedness and anguish, and the tormented din of thousands of birds shrieking their pain together, you lay an egg and watch it roll out of sight; but the joy of making a nest, of giving birth, of clucking to your chicks is absent—laying the egg is an empty, frustrating, and exhausting ritual (Coats, 93–94).

Most of the eggs sold for human consumption are infertile. Battery-caged hens do not have contact with cockerels except for those missed at the hatchery, where the male chicks are trashed—a quarter of a billion birds born each year in the United States—representing half the population of egg-industry hatchlings. They can't lay eggs or compete with "broiler" chickens for muscle tissue (meat); hence "the sex is terminal" (North and Bell, 4). Male chicks who escape the chicken sexers, and are not subsequently culled in the pullet house, where the young hens live in cages until they are transferred to the layer cages where they'll spend the rest of their lives, can end up caged with the hens. On my tour of a caged layer facility in Maryland, two or three crows rose amid the cries of thousands of hens.

Embryonic Development and Hatching of the Chick

If the egg is fertile, then a chick is beginning to grow inside, having already developed by the time it is laid from a one-celled individual, or zygote, to an embryo composed of approximately 60,000 cells through geometric cellular division. Total incubation takes 22 days, including one day inside the hen. While still in her body, the reproductive cell, the blastoderm, forms into two layers, an upper layer of cells called the ectoderm and a lower layer of cells called the entoderm. Soon, the middle layer, or mesoderm, forms. The entire body of the bird arises from these three layers. From the ectoderm comes the nervous system, parts of the eyes, the feathers, beak, claws, and skin. From the entoderm come the respiratory and secretory organs and the digestive tract. From the mesoderm come the skeleton, muscles, blood system, reproductive organs, and excretory system.

The mammalian embryo develops inside the mother from nutrients derived from her blood supply. The avian embryo develops outside the mother from nutrients that were made or received by her body and stored in the egg. Certain membranes make these nutrients available to the growing chick. The yolk sac that envelops the yolk secretes an enzyme that changes the yolk into a digestible form.

Just before hatching, the yolk sac is drawn into the chick's body to serve as a food supply for the first few days after hatching. The allantois, which completely surrounds the embryo by the ninth day, oxygenates the blood and removes carbon dioxide, removes excretions to a special cavity, and aids in the digestion of albumen and the absorption of calcium from the eggshell. The chorion fuses the allantois with the inner shell membrane to facilitate these functions. On the twelfth day, the embryo begins to imbibe the amniotic fluid, which may contain chemicals that stimulate the development of taste and smell (Rogers 1995, 22, 39).

A major challenge of incubation is the conservation of fluids and the preservation of the chick from harm, while allowing for the diffusion of oxygen into the egg and the release of carbon dioxide and moisture into the atmosphere. This challenge is met by the existence of thousands of tiny pores in the shell, backed up by cuticle that stretches across the pores to hinder evaporation and prevent bacteria from entering the shell. Organisms that manage to get through face disruption by the antibacterial protein lysozyme in the albumin.

The hen donates parental immunity to the developing embryo. Antibodies produced by her body in response to pathogens (disease-causing microorganisms) in the immediate environment pass into the egg, to be incorporated into the blood of the chick. When the hen and her chicks share the same environment, the unborn chick is thus protected from the very bacteria, viruses, and fungi that are most likely to cause trouble. However, parental immunity wears off soon after the chick hatches. Half is lost in the first three days. By the end of the fourth week, it disappears (North and Bell, 758–759). Under ordinary circumstances, this is no problem, for by then the chick has developed an active immune system, aided by a rich intestinal microflora that destroys and repels harmful invaders (Fowler, 5, 9).

Maternal Immunity Disrupted by Factory Farming: Marek's Disease and Infectious Bursal Disease

Even under conditions in which an organism and its environment are in harmony, diseases will sometimes occur. Normally, an organism's natural defense system holds diseases in check. However, factory farming, with its inherent filth, has produced specific diseases that penetrate parental immunity and disrupt the chick's developing immune system by attacking the two lymph glands in which immunity originates, the thymus (T-system) and bursa of Fabricius (B-system). Impairment of these glands disrupts the production of antibodies, reducing or eliminating the bird's ability to resist bacterial infections such as *Salmonella* and *E. coli.* Two examples are Marek's disease and infectious bursal disease (Bell and Weaver, 506–508).

Marek's disease is an infectious, immunosuppressive cancer that fills the chicken's spinal cord and peripheral nervous system with malignant tumors, resulting in paralysis, blindness, and death. It is caused by an airborne herpesvirus that localizes in the cells lining the feather follicles, where it replicates, and is sloughed through the dander and feather particles to float in the air and be inhaled by the birds in the dirty, crowded buildings in which even their feathers are fatal.

To control Marek's disease, the poultry industry vaccinates chicks at the hatchery with subcutaneous injections in the nape of the neck, or it injects the embryo by machine on the 18th day of incubation (Bell and Weaver, 519). According to Hunton, Marek's disease vaccines "appear to work by preventing development of signs of Marek's disease. However, they apparently do not prevent infection of the hosts with the Marek's disease virus: [the] virus can be recovered from almost any commercial chickens tested" (8). A 1996 article in *Broiler Industry* magazine stated: "Marek's disease condemnations and mortality are on the rise, increasing with each passing year during the 1990s" (Miller 1996, 30). In 2006, the USDA's Agricultural Research Service reported that because the virus is constantly evolving, "new vaccines have to be developed to keep them in check" (Durham).

Infectious bursal disease, also known as Gumboro, because the first outbreaks were reported near Gumboro, Delaware, in 1957 (Brooks, D1), is "an acute, highly contagious viral infection of young chickens that has lymphoid tissue as its primary target with a special predilection for the bursa of Fabricius" (Calnek, 648). There, it destroys the immune cells responsible for most of the antibodies in the young chicks, making them vulnerable to everything from skin disease to hepatitis to anemia to *E. coli* infections.

Afflicted birds develop severe liver and kidney disease and are listless, nervous, sleepy, and dehydrated, and have a whitish diarrhea. Their irritated vents cause the birds to pick at themselves. Filthy houses and equipment "are definite sources of infection" (Bell and Weaver, 507).

Inside the Egg

Meanwhile, a tiny being is growing inside the egg, whether nestled beneath the mother hen or crammed in an incubator among thousands of other embryos. During the first 24 hours after the egg is laid, the tiny heart starts beating and blood vessels begin to form, joining the embryo and the yoke sac. The nervous system originates during the 21st hour of incubation, followed by origination of the head and the eyes (North and Bell, 49–50). Other body parts begin to develop during this time, including the alimentary tract and the spinal column. On the third day, the embryo begins to rotate to lie on its left side. By the fourth day, all of the body organs are present, with the nose, legs, wings, and tongue taking shape and the vascular system in place.

On the fifth day, the reproductive organs differentiate and the face begins to assume a lifelike appearance. On the sixth day, the beak and the egg tooth (a kind of rough edge that disappears after hatching, which protects the beak and also helps crack the shell) can be seen, along with some voluntary movement of the embryo.

During the next seven days, the body develops rapidly, including the formation of the abdomen and intestines. Feather germs, the origin of feather tracts, appear, the beak begins to harden, toes and leg scales start to show, the skeleton begins to calcify, and chick down appears. On the fourteenth day, the embryo, now covered with down, rotates to arrange itself parallel to the long axis of the egg, normally with its head toward the large end of the egg near the air cell. On the seventeenth day, the chick turns its head, placing its beak under its right wing toward the lower part of the enlarged air cell to prepare for hatching and breathing outside the shell.

Hatching

On the nineteenth day, the yolk sac begins to enter the body through the umbilicus, and the chick positions itself for pipping the shell, that is, for making a hole in the shell to breathe through while struggling to get out. On the twentieth day, the yolk sac completes its absorption into the body cavity and the umbilicus begins to close. By now, the chick occupies the entire area within the shell except the air cell, which it now begins to penetrate with its beak, inhaling outside air through its lungs for the first time.

After pipping the shell to reach the air cell, the chick rests for several hours. It then cuts a circular line counterclockwise around the shell by striking the shell with its egg tooth near the large end of the egg, aided by a special pipping muscle in its neck which helps it to force its beak through the membranes lining the shell. With the egg tooth the chick saws its way out of the shell, aided by the mother hen if she is there and help is needed. Between 10 and 20 hours after the shell is first broken, the chick emerges, wet and exhausted, to face the life ahead.

"The split opened, tearing the inner membrane with it and spilling the first chick out on the straw where she sprawled, naked-looking, streaked with wet down. Granny Black picked the last pieces of shell gently from her, slaying every ant that darted in to attack, until those left alive had to content themselves scavenging off the sticky remnants of shell" (Gage, 2).

Nearly two days may elapse between the hatching of the first chick and the appearance of the last member of the brood. Thus, some chicks may be almost two days old by the time all of their sisters and brothers have struggled from their shells, as many as 16 others. However, hatching is not a haphazard process.

About 24 hours before a chick is ready to hatch, it starts peeping in its shell to notify its mother and siblings that it is ready to emerge (Rogers 1995, 41, 47–53). A communication network is established among the chicks, and between the chicks and their mother, who must stay composed and unruffled while all the peeping, sawing, and breaking of eggs goes on underneath her. "During all this time the chorus of peeps goes on virtually uninterrupted, the unborn chicks peeping away, the newborn ones singing their less muffed song" (Smith and Daniel, 321). Since some of the eggs may be infertile or aborted, the peeps tell the hen how long she needs to continue sitting on the nest.

Mother Hen and Chicks

As soon as all the eggs are hatched, the hungry mother and her brood go forth eagerly to eat, drink, scratch, and explore. Baby chicks are precocial, meaning they are genetically equipped to find food and follow their own kind, or whoever is in charge, in the process known as imprinting. Imprinting, as Lesley Rogers explains, is "a powerful form of learning by young chicks and ducks, as well as by other species that are born in a relatively advanced state of development. By the process of imprinting they learn to recognise their mother and so follow her" to ensure their survival. By imprinting, the "chick learns the features of the hen and also of its siblings, and it remembers these for a very long time, possibly for the rest of its life" (Rogers 1997, 62–63, 73).

The chicks practice hygiene by preening and dustbathing almost immediately. However, their primary dependency is the need to stay warm and dry.

> Periodically the mother squats down, perhaps alerted by some change in the decibel of her chicks' peeps—a peep, say, that indicates they are chilled and in need of warmth—and they all dash under her outspread feathers and stay there until they are thoroughly warmed; then out again to continue the search for food and the adventure of exploring the world (Smith and Daniel, 323).

The chicks venture fairly far away from their mother, communicating back and forth all the while by clucks and peeps. The hen keeps track of her little ones on the basis of color, possibly also by smell, and by counting the peeps of each chick and noting the emotional tones of their voices (Smith and Daniel, 323–324; Rogers 1995, 83–84). Should a peep be missing or sound frightened, she runs to find the chick and deliver it—not always successfully—from the hole in the ground, tangled foliage, or threatening predator.

During the first four to eight weeks or so, the chicks stay close to their mother, gathering beneath her wings every night at dusk. Eventually, she flies up to her perch, indicating her sense that they, and she, are ready for independence (Smith and Daniel, 324). Young chicks without a mother huddle together at night for the first month or two. Then one evening you see them practicing sitting in a row, before huddling. Then comes an evening when they are lined up on their perch, arranging and rearranging themselves as before, only this time they stay lined up all night, henceforth roosting at night like adults.

Commercial Hatchery

First-class chicks are delivered to the farms. Second-class chicks are sold at a reduced price to individual customers. Third-class chicks (severe problems) have no value and are destroyed.

—Bell and Weaver, 782

Chickens waking up in a commercial hatchery have a totally different experience from chicks hatching under a mother hen. A former Merck pharmaceutical company employee described her introduction to this world:

> My first hatchery tour came the next day. For the uninitiated, the hatchery is the place where chicken eggs are incubated in large walk-in incubators. Everything is timed so that on the prescribed day a particular incubator is opened and most of the eggs have hatched fluffy yellow chicks. The huge wheeled carts inside are rolled out and wheeled down the hall to the waiting window, much like the ones found in school cafeterias where students return their lunch trays. Next to the window were three workers. It was their job to remove each tray of newly hatched chicks from the cart, pick out the live chicks and toss them through the window onto a conveyer belt, and then dump the discarded shells into the trash. They did this very quickly. In fact, so quickly that often the conveyer belt would get backed up with the chicks and they would have to stop cleaning off the trays and wait. The men used this time to puff on their cigarettes or just stand there. This would not have bothered me if I had not noticed an overly energetic chick hop onto the edge of the tray and fall onto the floor. The workers ignored the chick and continued smoking. As my eyes followed the chick's descent, I realized that he was but one of many to make that trip. Although they landed apparently unharmed, they did not stay that way.
>
> As soon as all the trays of chicks had been removed from the cart, it was wheeled away, smashing several escaped chicks as it went. The ones that managed to miss being run over by that cart were the prime target of the next cart's wheels. I looked around the floor—it was littered with smashed and half-smashed chicks. Some were trying to move, but couldn't overcome the gluelike hold of their smashed, blood-soaked wings. I had to look away and pretend not to notice. I felt that in this situation there was nothing I could say or do that would make any difference (Farb 1993a).

Extremely large hatcheries produce two million or more chicks every week (Bell and Weaver, 700). Hatcheries producing the "egg-type" hens for laying the eggs people buy at the store destroy the unwanted male chicks—approximately a quarter of a billion male chicks in the United States alone each year. Methods of destroying the unwanted male chicks, and the defective and slow-hatching chicks of both sexes, include throwing them in trash cans to suffocate to death, grinding them up alive (maceration), electrocution, and gassing them with carbon dioxide. Collectively, these "cull" chicks are "hatchery waste" to be landfilled or turned into pet food and farmed-animal feed (Bell and Weaver, 793–794).

In the hatchery's "chick processing room," male chicks and some female chicks intended for breeding purposes have their toes cut off at the outer first joint of the back toe of each foot. Male chicks intended for breeding may also have the outer first joint of the inside toe, or of two inside toes, of each foot removed with an electric toe clipper (North and Bell, 251). The combs (the red crests on the heads of roosters and hens) are dubbed (removed) by running a pair of manicuring scissors or shears from the front to the back of the comb close to the head of the male chick, and spurs on young chicks intended for breeding may also be trimmed off. "Use an electric beak trimmer or toe clipper to remove both the toes and spurs," *Commercial Chicken Meat and Egg Production* advises, cautioning that this "very delicate procedure" causes an additional toll of injuries and deaths (Bell and Weaver, 625, 1001).

Comb dubbing is said to be best done when the chicks are a day old to avoid the severe hemorrhage that is likely to occur after the first day. Producers are advised not to dub birds in warm climates, as the comb functions to eliminate excess body heat (North and Bell, 251–252).

Male chicks intended for breeding, and female chicks used by the commercial egg industry for both breeding and table-egg production, are debeaked ("beak trimmed") at the hatchery, or shortly after being trucked to the grow-out facility. Because they are slaughtered as six-week-old babies, day-old "broiler" chicks are no longer debeaked (Bell and Weaver, 701).

Though beak trimming can be done in several ways, "the most common method uses a sharp, hot blade to facilitate both a clean cut and simultaneous cauterization of the wound," according to *Commercial Chicken Meat and Egg Production*, which warns that because beak trimming "is a major stress," the procedure should be separated from "other similar stresses by at least two weeks" (Bell and Weaver, 1001). Elsewhere it says that "male breeder chicks are beak-trimmed and de-toed at the hatchery," and since young beaks often grow back, "birds may have to be re-trimmed" (Bell and Weaver, 701–702).

An undercover investigator employed in the early 1990s by Hudson's hatchery and breeder farms on the Eastern Shore of Maryland said that after a few weeks the company let him debeak chickens. In the process of having their beaks burned off, the birds chirped loudly and defecated profusely: "In pain, these birds flap their wings, push against the machine, and often lose control of their bowels," the investigator wrote. The stench was terrible: "Smoke rises from the place where the beak meets the machine as the bird loses at least an eighth of an inch of her beak. A few inches higher up, another part of the machine cauterizes her wound. Because of the speed at which the workers handle the chicks, 'hack jobs' result in massive beak loss to some chicks, leaving them unable to eat." Many birds die within 24 hours from shock and blood loss.

As noted, many chicks are debeaked not once but twice, due to beak regrowth and other problems. Improperly cauterized birds bleed from their wounds. "'Bleeders' are easy to recognize by the spots of red down their fronts or under their wings where the birds have tried to preen," the Hudson investigator noted. "Some days there were more than 100 bleeders." One morning he found 42 chicks who had bled to death overnight.

Chicks are vaccinated at the hatchery against Marek's disease and other contagious diseases by a combination of mechanical injectors, vaccine sprays, and manual syringes. Manual vaccination is ugly. Workers handling 2,000 to 3,500 birds per hour—up to 8,000 birds per day—grab baby chicks and hold them while an automated vaccination needle punctures the back of their necks" (Smith 1996; Bell and Weaver, 701).

Vaccination is a primary cause of infection in the young birds. The puncture breaks and may even tear the skin during the rapid process, and the same needles are used over and over, spreading contamination ("Immune system," 26). The foreman at Seaboard Farms hatchery in Mayfield, Kentucky, told a visitor that, in addition, "the chicks can suffer spinal injury or brain damage if vaccinated in the wrong area of the neck" (Danowski).

Automated egg injection systems are designed to inoculate 20,000 fertile eggs an hour against Marek's disease on the 18th day of the 21-day incubation period. They are favored by the big poultry companies to reduce manual labor and carpel tunnel syndrome among hatchery workers as well as the harsh handling of newly hatched chicks, which stunts the birds' growth rate.

Among the vaccination systems described in *Commercial Chicken Meat and Egg Production* is one called the "robotic automatic vaccinator," which performs debeaking and vaccination in a single process "using a laser needle and spray vaccine." The chick's beak is placed in a hole, and the machine automatically debeaks and vaccinates approximately 4,300 chicks per hour (Bell and Weaver, 701).

Treatment of Parent Flocks

Chicks destined to serve as parent flocks are injected again in the breeder houses. Workers at the Hudson's breeding facility in Maryland "catch three birds at a time and hold them by both wings held together behind their backs to expose the chest and wing 'pit,' where injections are administered." As a bird is lifted for an injection, "her squawking becomes faster and higher and takes on a frantic tone." Injecting birds himself, an investigator said he "could definitely feel things inside the wings snapping." Workers used the same needle on one chicken after another.

Not surprisingly, many employees vent their frustrations on the birds. A supervisor swore "about a bird who had escaped during unloading and eluded capture, [so he] threw a board at her and missed, then kicked her four or five feet into the air. Another shouted obscenities at a chicken he blamed for having made him fall and twist his ankle, [he] then lunged at her, throwing his whole weight on her, and punched her twice. Another, who broke the wing of a bird causing a bone to protrude, blamed the chicken for not letting him catch her" ("Chicken Hell," 11–12).

Why Look at Chickens

Perhaps if we had realized they are birds, with all the wonderful characteristics of birds, we would have paid closer attention to the ways in which chickens can enchant us.

—Jeffrey Masson, *The Pig Who Sang to the Moon*, 59

This is the world that we have made for chickens to live in. Some people feel threatened by the prospect that in recognizing and upholding the dignity of other living beings, we betray our own dignity as a species. It should rather be asked how the human species gains dignity by creating worlds such as this for anyone to live in. Can one regard a fellow creature as property, an investment, a piece of meat, an "it," without degenerating into cruelty and dishonesty toward that creature? Human slavery was brutal. Does anyone really believe that nonhuman animal slavery operates on a higher plane?

We need to change. Let us change. Let us begin to see chickens, and the world we share, with more envisioned eyes. Alice Walker wrote of her experience:

> It is one of those moments that will be engraved on my brain forever. For I really *saw* her. She was small and gray, flecked with black; so were her chicks. She had a healthy red comb and quick, light-brown eyes. She was that proud, chunky chicken shape that makes one feel always that chickens, and hens especially, have personality and *will*. Her steps were neat and quick and authoritative; and though she never touched her chicks, it was obvious she was shepherding them along. She clucked impatiently when, our feet falling ever nearer, one of them, especially self-absorbed and perhaps hard-headed, ceased to respond (Walker, 171).

Whenever I tell people stories about chickens enjoying themselves, many become very sad. The pictures I'm showing them are so different from the ones they're used to seeing of chickens in a state of absolute, human-created misery. Many people are amazed to learn that chickens have personalities. *The New York Times* restaurant critic William Grimes wrote of a beautiful black hen who entered his life unexpectedly one day, an apparent escapee from a live poultry market in Queens. "I looked at the Chicken endlessly, and I wondered. What lay behind the veil of animal secrecy? Did she have a personality, for one thing?" His curiosity is satisfied by close acquaintance with and observation of the endearing bird. By the end of his bittersweet book *My Fine Feathered Friend*, he and his wife Nancy "had grown to love the Chicken" (59, 81).

This is why we have to start looking at chickens differently, so that we may see them, as Alice Walker says she "absolutely *saw*" the Balinese chicken crossing the road one day with her three little chicks. She explains that having perceived the being of this particular chicken, she can never again not see a chicken. Her obligation to "a sister, [whose] love of her children definitely resembles my love of mine," starts with this moment of vision.

CHAPTER 3

THE LIFE OF THE BATTERY HEN

Please, never, ever, call me a battery hen. . . . A hen in a battery cage is not a battery hen. . . . I, Minny, am a proud descendant of the Red Jungle Fowl.

—Clare Druce, *Minny's Dream*, 2004, 48–49

We have followed the laying hen from the trees to the barnyard and through a multitude of types of laying quarters to today's buildings that are fully automated as well as light and temperature controlled. We have seen how chore time for feeding, watering, cleaning, and gathering of eggs has been greatly reduced by improved technology. Old practices such as routine culling of nonlayers are no longer followed, and new practices like forced molting have been introduced.

—Wilbor O. Wilson, "Housing Environment has Been Man's Concern Ever Since the Hen's Jungle Days," 1974, in Skinner, 244

Here the birds are locked up, trapped twice, doubly trapped. First you are in a cage, and then you are in this building loaded with all these pathogens and flies and toxic gases and lights burning in a stinking, cobwebby gloom. There is an endless sound of machines and distressed birds all around you. You can't even describe it to people. What we need in addition to video footage is for something to enable people to smell what it is like in there. These birds are creatures with wings and legs. To take creatures with wings and legs and never let them take a step is horrible.

—Karen Davis, *Egg Industry*, cited in Olentine, October 2002, 18

In no way can these living conditions meet the demands of a complex nervous system designed to form a multitude of memories and to make complex decisions.

—Lesley J. Rogers, *The Development of Brain and Behaviour in the Chicken*, 1995, 218

Egg laying in birds is a biological activity based on the ingestion and absorption of a specific combination of nutrients in the presence of light. A hen knows how to select the calcium and other nutrients she needs (North and Bell, 658, 673). Her sense of the length of the day enables her to synchronize her laying cycles with the cycles of nature. Sunlight passes into her eye, sending a message to her brain, which in turn passes its own message to the anterior pituitary gland, which produces a hormone that causes the ovarian follicle to enlarge. The ovary generates the hormones, or sex steroids, that stimulate the processes required to form an egg.

Estrogen sparks the development of the medullary bone for calcium and the formation of yolk protein and lipids (fat) by the liver. It increases the size of the oviduct, enabling it to produce albumen proteins, shell membranes, calcium carbonate for the shell, and the shell cuticle, which acts as a lubricant during the laying process and subsequently as a dry shield against bacterial penetration of the egg once it has been laid.

Caged-laying-hen producers artificially stimulate and extend egg production by keeping the lights burning for 15 to 16 hours a day, mimicking the longest days of summer, to force the pituitary gland to secrete increased quantities of follicle-stimulating hormone, which in turn activates the ovary (Bell and Weaver, 1038–1039). The wife of a management team at Country Fair Farms in Westminster, Maryland, told us during a tour, "It is hard on the hens. Don't think we don't know this" (Fleishman).

Commercial laying hens are not bred for their flesh, so when their economic utility as egg layers is over, the still young birds are disposed of as cheaply as possible. Some are suffocated to death in 40-foot-long dumpsters, then trucked to rendering plants and turned into animal feed ingredients. Others are gassed and buried dead or alive in landfills, or ground up alive in woodchippers. Still others are trucked to slaughterhouses and turned into meat products such as "spent hen meal," which is fed back to the hens as a feed ingredient. In compiling its annual poultry slaughter statistics, the U.S. Department of Agriculture's National Agricultural Statistics Service lumps together hens used for commercial egg production with roosters and hens used for breeding purposes and divides them into two categories of mortality: "Lost" and "Sold." "Lost" includes chickens who were "rendered, died, destroyed, composted, or disappeared for any reason except sold." In 2004, 101,079 of these types of chickens were recorded as "lost" and 192,066 were recorded as "sold for slaughter" (USDA-NASS 2005, 5–6).

In the process of being pulled from the cages and stuffed into transport crates, "spent" laying hens are treated even more roughly than meat-type chickens because of their low market value. Their bones are very fragile from lack of exercise and from calcium depletion for heavy egg production, causing fragments to stick to the flesh during processing. The practice known as forced molting, in which hens are deprived of all food or placed on nutritionally deficient diets to regulate the price of eggs, results in beaded ribs that break easily at the slaughterhouse. Removal of food for several days before the hens are loaded onto the truck weakens their bones even more (Brown 1993).

Both the U.S. egg industry and the American Veterinary Medical Association oppose humane slaughter legislation for laying hens, claiming that their low economic value does not justify the cost of "humane slaughter" technology. Nothing has changed since the AVMA wrote to me in 1992 that "spent fowl (culled laying hens) frequently suffer broken bones when electrically stunned, and their low economic value (.02–.08/lb) makes it difficult to justify costly new slaughter techniques" (Boyce).

First Impressions

It is difficult to convey to anyone who has not witnessed it directly the mistreatment of laying hens that began in the twentieth century. Page Smith and Charles Daniel state in *The Chicken Book* that the rows upon rows of birds, with their mutilated beaks, in the small cages, are "like a glimpse into an Inferno as terrible in its own way as any of the circles of Dante's hell" (287). Lynn Shepherd, a college student, described her tour of the Milton Waldbaum egg farm in Colorado:

> Joy [the manager] was explaining various functions of the different machines as I was eagerly searching for a glimpse of the thousands of chickens I could hear squawking. Soon we rounded the corner and all my fears came to life. The appearance of the chickens was like all the horrible pictures I had discovered in my research, those pictures that I had thought were so exaggerated.

> The chickens were without any neck feathers and their necks were covered with blisters. Their wings were bare with an occasional half feather extended from them. The second cage I looked at contained a dead, purple, featherless carcass on the bottom of the cage. Joy explained that the chicken house attendant must have missed this one when he did his morning rounds.

A woman who was thinking of starting a backyard chicken flock purchased four "spent" hens from a commercial farm in Massachusetts. She wrote:

> My own first impression was "these are not chickens, oh no they aren't." In front of me in the dust lay (not walked, lay) four small bodies, barren of the feathers I knew that nature had endowed to chickens. Toenails were four and a half inches long. Words like "pathetic" and "sick" and "ravaged" leapt to mind and then to tongue. In all our reading about raising chickens, in all those homesteading books and Extension Service pamphlets, never had we seen described what life on the egg farm does to the body of the hen. Now we know (Costello).

The egg industry misleads the public to believe that concern for the suffering of hens is mere sentimentality. Indeed, "Anyone who is in a layer house early in the morning will hear the chickens sing. They are comfortable; they are happy" (Wentink). During my tour of Country Fair Farms, a battery hen facility in Maryland, in which 125,000 debeaked hens lived nine to a 20 x 20-inch cage amid constant raucous, shrieking noise, the manager said that "the sound we heard was singing. The hens sing in their cages."

A contributor to *Commercial Chicken Meat and Egg Production* compares the location call of wild and feral chickens, which consists of a "gentle, drawn-out, low-frequency vocalization often interspersed with several shorter calls of similar intensity, to maintain contact with other group members in the underbrush," with the "singing" of hens commonly heard in commercial layer houses (Bell and Weaver, 72).

An industry that calls people who claim that caged hens are miserable "anthropomorphic" has no problem anthropomorphizing these birds as contented songsters.

The Cage System of Egg Production

Large farms of one million or more laying hens have become commonplace in the United States. . . . [L]arge farms are necessary to keep equipment operating at capacity.

—Bell and Weaver, 947.

Egg producers use higher densities to increase total production and to reduce overhead costs per dozen eggs.

—Bell and Weaver, 1018

The modern hen laying eggs for human consumption is far removed from the tropical jungle fowl from whom she originated and from the active farmyard hen of a few generations ago. The battery-caged hen is an anxious, frustrated, fear-ridden bird forced to spend 10 to 12 months or longer squeezed inside a small wire cage in the midst of a building full of cages suspended over manure pits. The cages are set in rows of tiers from two to ten stacks high, in 500- to 600-foot-long buildings (152–183 meters), like a factory warehouse full of boxes.

In 2003, *United Egg Producers Animal Husbandry Guidelines for U.S. Egg Laying Flocks* estimated that "98 percent or more of the commercial egg production in the U.S. and an estimated 70–80 percent of the world's egg production are derived from caged layers." The *Guidelines* predicted that despite a trend away from cages in some European countries, "the increasing use of cages in developing countries will continue to increase the percentage and population of layers housed in cages worldwide."

In 2004, the International Egg Commission listed 15 countries with 95 percent or more hens in battery cages: Argentina, Brazil, Canada, Chile, Cyprus, Iran, Italy, Japan, Kazakhstan, Mexico, Portugal, Russian Federation, South Korea, Spain, and the United States ("Industry News"). The 2006/07 WATT *Executive Guide To World Poultry Trends* forecasts global egg consumption and production to increase, with the current number of laying hens worldwide said to be 5,646 million birds in 2005, or 31 percent above the 1995 level (28, 30).

Brief History of the Battery Cage and Criticism

An article in *Egg Industry* magazine in 2006 tells the history of U.S. egg production over the past 50 years (Rieger). Prior to the conventional use of cages in the 1960s, flocks of 250 to 10,000 birds were kept in floor-type houses where feeding, watering, and egg collection were done by hand. In the late 1960s, a "great cage debate" arose among breeding companies, poultry professors, universities, building designers, and equipment manufacturers over what types of facilities to invest in for the future. The cage system prevailed, with the result that by the early 1970s, new buildings were being designed for flocks of 10,000 to 30,000 hens. In the late 1970s and 1980s, larger buildings were constructed to accommodate 50,000 to 120,000 caged hens per building.

In the 1980s, the egg industry began moving away from single-standing caged layer houses toward in-line complexes "where you could put a million layers or more on a single site, then connect the houses by a common corridor and an egg belt, with all the egg production flowing into a single processing and packing plant." Switching to these in-line, multibuilding complexes led to facilities that can hold "from 150,000 to 380,000 layers in a single house with 2 to 5 million birds on a single site" (Rieger, 6).

In the United States, a three- to four-pound hen requiring 74 square inches merely to stand, with a wing span of 30 to 32 inches (Eleazer), may be legally confined with five to ten other hens in a cage that is 15 to 16 inches high, 12 inches deep, and 18 to 20 inches wide, depending on the number of hens in the cage (Bell and Weaver, 1009). Prior to 2003, each hen had an average living space of 48 square inches, and recent undercover investigations indicate that many, if not most, still do. In the early 1980s, the U.S. egg industry trade group, United Egg Producers (incorporated in 1968), published its first *Recommended Guidelines of Husbandry Practices for Laying Chickens*, stating: "Irrespective of the type of enclosure or system of management used, all birds shall have sufficient freedom of movement. An average of 48 square inches per bird or 12 square inches per pound of bird liveweight is adequate" (UEP, n.d.).

Criticism of the cruelty of the battery-cage system developed in Switzerland, Sweden, and the United Kingdom in the 1970s and spread to the United States in the 1980s and 1990s. Michael Baxter summarized the criticism in the *Veterinary Record* in 1994, noting that the "space available in a battery cage does not allow hens even to stand still in the way they would in a more spacious environment" (Baxter, 617).

By the late 1990s, the "growing concern for animal welfare worldwide" prompted United Egg Producers to commission a Scientific Advisory Committee for Animal Welfare in 1999 "to develop recommendations based upon existing science for presentation to the UEP Producer Committee for Animal Welfare and ultimately to the industry." This was explained in UEP's *Animal Husbandry Guidelines for U.S. Egg Laying Flocks 2000 Edition*. Updated in 2003 and again in 2006, the *Guidelines* set a goal of 64 square inches for each White Leghorn hen and 72 square inches for each brown egg-laying hen by 2007 for chicks hatched after October 1, 2006. The final goal, set for 2008, is 67 square inches for each White Leghorn hen and 76 square inches for each brown laying hen. To meet this goal, equipment purchased, contracted for, or built after April 1, 2003, and installed by December 31, 2003, must accommodate the placement of hens after August 1, 2008, at a minimum of 67 square inches per hen for White Leghorn hens and 76 square inches for brown egg layers. The *Guidelines* set "optimized" welfare standards in the range of 67 to 86 square inches of usable space per bird.

Laying Eggs in Cages

In the twentieth century, the combined genetic, managerial, and chemical manipulations of the small and lively Leghorn chicken of Mediterranean descent produced a bird capable of laying an unnatural number of eggs—an average of 223 eggs per year (Bell and Weaver, 949), in contrast to the one or two clutches of about a dozen eggs per clutch laid in a year by her wild relatives (Druce 2004, 41).

Genetic selection for early egg laying to reduce the time and money required to feed and house growing birds for six months can result in the formation of eggs that are too big to be laid by the immature body of a small, five-month-old bird. Uteruses can prolapse, pushing through the vaginas of the small, cramped birds forced to strain day after day to expel large eggs. The uterus protrudes, hangs, and "blows out," inviting infection and vent picking by cell mates, from whom the prolapse victim, in severe pain, cannot escape except by dying. A small-farm handbook links a prolapsed uterus to the hen's "inability to stand the constant exercise of egg laying, often promoted to excess by use of chemical stimulating laying meals, salts, powders, etc., and the cruel method of prolonging laying hours by use of artificial lighting after the normal dusk roosting" (Levy, 300).

The laying of an egg has been degraded by the battery system to a squalid discharge so humiliating that ethologist Konrad Lorenz compared it to humans forced to defecate in each others' presence. In his article “Animals Have Feelings,” Lorenz wrote, “Everyone knows what a battery hen looks like. Bloody combs, misshapen claws, etc. There has been much debate over the frustration of the instincts of such battery hens. This has been proved beyond the shadow of a doubt by the limitation of mobility, the beating of the wings. . . . The animal expert knows what a terrible sight it is to see a hen trying time and again to crawl under the other hens in order to find cover and protection. There is no doubt that hens in these conditions tend to delay laying their eggs. Their hesitation to lay their eggs in the close neighborhood of the other cage inmates is just as instinctive as the hesitation of a civilized person to defecate in front of others in a similar situation.”

A free hen chooses her nesting site carefully and prepares her nest in a purposeful manner, known as the laying ritual. During this time, she composes herself to enable her body to conduct the intricate processes that culminate in the laying of the egg, as described in chapter 2. As Lorenz noted, a hen will normally wait to lay her egg until she is comfortably settled in her favorite place. At our sanctuary, all of our hens have their favorite nesting areas, including an oval niche amid a pile of books in the cellar made by our hen Charity. If the cellar door was closed, blocking her steps to the basement when she was ready to lay her egg, she would pace back and forth in front of the window on the opposite side of the house where I sat at my desk. When I opened the door, she ran down. There was no mistaking her meaning.

Poultry researchers have described the futile attempts of caged hens to build nests and their frantic efforts to escape the cage by jumping at the bars right up to the laying of the egg. Deprived of nesting materials, a caged hen will stereotypically “pace” in the square inches allotted her instead of sitting quietly on the nest (Vestergaard 1987, 10). Finally, the egg drops from her body onto the wire cage bottom, then on to the conveyer where it rolls out of sight.

Diseases and Syndromes

Disease and suffering are inherent features of the battery-cage system in which the individual hen is obscured by gloom and by thousands of other hens in an environment designed to discourage perception, labor, and care. The lack of care is characterized in *Commercial Chicken Meat and Egg Production*, which boasts: “Today’s complexes are able to reduce house caretaking and egg packing requirements to 15 persons per million hens.” Only five “house caretakers” are needed for “10 layer houses with a capacity of 105,000 birds each” (Bell and Weaver, 956, 974, 977–978).

Forcing a physically active bird to assume a cramped and stationary position for life on wire mesh produces diseases that are complicated by abnormal reproductive demands: muscle degeneration, poor blood circulation, accumulation of flaccid fat, oviducts clogged with masses and bits of eggs that can't be expelled, osteoporosis, and foot and leg deformities (Calnek 1991, 827–862). The filth of the debeaking machines, vaccination equipment, and overall living conditions has produced an incurable disease in laying hens known as swollen head syndrome.

Swollen Head Syndrome

This infectious syndrome, also known as cellulitis, attacks birds used for breeding and for commercial meat and egg production in intensive confinement systems. The bird's face puffs out as a result of the swelling of the layers of cellular tissue beneath the skin, which is full of pus underneath. Swollen head syndrome is accompanied by egg peritonitis, mucus congestion, nasal discharge, and nervous signs called cerebral disorientation. The disease has been identified with the turkey rhinotracheitis pneumovirus—another disease of confinement—and with a variety of bacteria, particularly *E. coli.* Cellulitis, which can affect other body parts as well, is caused by poor hygiene. For this reason, it "will continue to be a concern of the North American poultry industry for the foreseeable future" (Norton, 32).

A rescuer of spent caged laying hens in Mississippi who encountered swollen head syndrome was told by the veterinarian who examined the hens, "Basically, what this translates into is that these birds were kept in filth. It wasn't necessarily the droppings, but just that the environment itself is filthy" (Stanley-Branscum 1996).

Foot and Leg Deformities

The feet and legs of chickens contain complex joints including many small bones, ligaments, cartilage pads, tendons, and muscles that enable the birds to search and scratch for their food on land that is rich in natural compost, insects, and plant life. Behavioral scientist Marian Stamp Dawkins explains that jungle fowl, "which are the wild ancestors of our domesticated chickens, spend long hours scratching away at the covering of leaves that hides one of their favorite foods—the minute seeds of bamboo. An ancestral memory of this way of life seems to have carried down the generations into the cages of our modern intensive farms so that even highly domesticated breeds have the same drive to scratch away to get their food—if they have the opportunity" (Dawkins 1993, 57).

Despite these facts, the battery hen spends her life standing and sitting on sloping, wire mesh rectangles designed to facilitate manure removal and the rolling of eggs onto a conveyer belt with minimum breakage. Her feet become sore, cracked, and deformed. Her claws, which evolved in nature to scratch vigorously, and thereby stay short and blunt, become long, thin, twisted, and broken. They can curl around the wire floor and entrap her, causing her to starve to death inches from her food and water (Appleby 1991, 7–8). When the chicken catchers wrench the hens from the cages at the end of the laying cycle, limbs, claws, and heads are frequently left behind, caught in the wires and trapped between bars, disputing the claim in *Poultry Behaviour and Welfare* that better cage design has reduced injuries and deaths (Appleby 2004, 148).

In experiments where hens are offered a choice between wire mesh floors and more natural materials, the wire mesh floor is "quickly abandoned in favor of floors consisting of peat, earth, or wood shavings." Dawkins goes on to explain that when hens who've been kept "all their lives on wire floors with no sight or contact with anything that could be scratched or raked over are suddenly, at the age of four months, given access to a floor of wood shavings or peat, even these naïve hens have an immediate and strong preference for these more natural floors over the wire ones, which is all they have known until then. They dustbathe, eat particles of peat, and scratch with their feet. It is not just the extra comfort afforded by a soft floor that attracts them, but all the behaviour they can do there as well" (Dawkins 1993, 153).

Caged Layer Hysteria, Fatigue, and Fear

A product of man's concentration of poultry under situations of stress is the appearance of a condition known as avian hysteria. The afflicted flock's behavior is typified by extreme nervousness, often exhibited at closely spaced time intervals with clocklike regularity. . . . This condition, first reported in the late 1950's, is linked to man's concentration of bird numbers in conditions of total confinement.

—Wilson quoted in Skinner, 234

Hysteria in Barren Cage Environments*. . . . R.S. Hansen of Washington State University carried out a series of experiments (Hansen, 1976) on the type of hysteria that may occur in laying hens, usually after they have been in cages for at least 100 days. . . . Hansen found the incidence of hysteria to be closely related to "population pressure"; with flocks of 40, 30, or 20 per cage, he had 91, 50, and 22% hysteria, respectively. Toenail removal reduced hysteria...*

—Craig, 246

Beak-trimming is known to help flocks with a hysteria problem.

—P.C. Glatz, 1

Physical contact with a human has been shown to elicit fear in hens. The level of fear experienced by the hens depends partly on the position of the cage in the poultry house and the bird's previous experience with humans. Birds from the upper tiers are generally more fearful of humans than are birds from lower tiers that have had more contact with humans.

—Kristensen, 175

The fact that hens are restricted from exercising to such an extent that they are unable to maintain the strength of their bones is probably the greatest single indictment of the battery cage.

—Michael R. Baxter 1994, 618

Caged layer fatigue is the term that is used to describe the condition of osteoporosis—loss of bone tissue—in laying hens kept in cages. It is characterized by an inability to stand, bone fragility, paralysis, and fractures. Affected hens have a washed-out appearance in their eyes, comb, wattles, legs, and feet. Disuse osteoporosis, resulting from lack of exercise, is exacerbated by secondary osteoporosis, the demineralization of bones and muscles for constant eggshell formation in hens specifically bred for egg production. According to *Diseases of Poultry*, "The condition has been restricted to birds kept in cages" (Calnek, 835).

Related to caged layer fatigue are hysteria and fear. Birds whose bones become paralyzed cannot reach their food and water. Videotapes show hens beating wildly against cage bars, shrieking, and hens with their heads and wings stuck between cage bars, their terror frozen in their faces and their eyes.

In 1981, Klaus Vestergaard cited 84 studies conducted between the 1940s and 1970s on the effect of cage systems versus non-cage systems on frustration, fear, and hysteria responses in laying hens. No matter how "flighty" the genetic stock was, cages produced the worst effects. For instance, fear and escape reactions occurred when an object was presented in front of caged hens, whereas in deep litter pens, reaction to an object "was rather that of curiosity and approach." The studies showed that "hens are more fearful in battery cages than in pens" and "the fear tends to increase with density. Hysteria, which is characterized by sudden wild flight, squawking (fear squawking?), and attempts to hide, has been interpreted as an abnormal fright-fear behaviour" (Vestergaard 1981b).

More than 20 years later, *Commercial Chicken Meat and Egg Production* talks about "emotionality (fearfulness, hysteria)" and fatigue in caged laying hens. It appears that "[g]enetic stock differences in flightiness due to fearfulness may have welfare implications for egg-type stocks of hens prone to osteoporosis. Exaggerated escape reactions could cause increased rates of bone breakage during flock removal," and various types of prelaying behavior, including pacing, sitting, and frustrated nest-building efforts, are performed in "repetitive, stereotypic fashion by egg-type stocks in cage systems, leading to the suggestion that some stocks experience frustration of nesting motivation in cage environments" (Bell and Weaver, 89–93).

The "Great Cage Debate"

Biologist David Sullenberger tells the history of caged layer fatigue:

> "Caged layer fatigue" syndrome became a real problem across the industry right after great quantities of birds began being confined in cages in the late 1950s. Until that time layers were floor birds much as broilers are today—with a big exception—and that is the layer houses had windows or skylights or sometimes both in the north, or were practically wall-less in the south and California. That is to say, the birds received a lot of sunlight even though much of it was reflected. Being chickens they also scratched around in the litter a lot and pecked at and ate a lot of what they found down there. Much of what they found was manure, or bits of food and litter with manure on it. . . . As a result, the birds were getting enough calcium, phosphorus, and magnesium back into their systems to supplement what they were being fed in their rations, which at the time, due to ignorance, were only marginally enough for good bones and high egg production. They were also getting sunlight, which means they were getting ultraviolet, which in turn meant vitamin D was being properly activated and calcium transport was very efficient.
>
> When cages came into vogue, ration formulations didn't change. The birds could no longer get the supplementation from the manure, eggshells came from the bones, and the bones became weaker and weaker. The smooth muscle tissue (heart, arteries) no longer had sufficient calcium ionic exchange because serum levels were low or out of balance

> or both. There were more heart attacks, and stress levels were elevated because birds were confined and the social order was completely messed up. Immune response became compromised, and bacterial and viral infections wiped out many birds.
>
> Caged layer fatigue syndrome became a virtual epidemic. Hundreds of thousands of birds across the nation were dying—virtually at the same age and stage of production. The industry panicked and nearly went back to floor houses. Would that they had. As you can imagine, a massive research effort was launched, the various problems surfaced, feed formulations were modified, vitamin D3 was added along with some other modifications to feed formulation, and "research once against triumphed over Nature" (Sullenberger 1994).

Caged layer fatigue continues. Bone weakness in caged hens is exacerbated by lack of exercise. Up to 30 percent of caged hens suffer broken bones when they are caught, transported, and slaughtered compared to "around half as many breakages in birds from free range or percheries" (Appleby 2004,148–49).

Fatty Liver Syndrome

According to *Diseases of Poultry,* "The shift from floor-reared layers to cage confinement was not without creation of new problems. . . . Among these is the fatty liver syndrome characterized by fat deposits, fatty livers, and drop in egg production." Fatty liver hemorrhagic syndrome is characterized by an enlarged, fat, disintegrating liver covered with blood clots, and pale combs and wattles covered with dandruff. "[S]mall hemorrhagic areas may be seen beneath the liver capsule. The liver is yellow, greasy, and of mush-like consistency. Large deposits of fat may line the abdomen and cover the intestines." The feeding of inappropriate high-energy diets to "birds whose exercise is restricted in cages" is cited as a possible cause of fatty liver syndrome in *Diseases of Poultry,* published by the American Association of Avian Pathologists, the veterinary arm of the U.S. poultry industry (Hofstad, 775; Calnek, 849).

SALMONELLA

These larger in-line operations with multiple houses . . . do not lend themselves to effective cleaning and disinfection.

—Charles W. Beard, DVM, "Our Industry Perspective on Research Needs for Salmonella Enteritidis"

In the last half of the twentieth century, hens' oviducts became infected with *Salmonellae* bacteria, which normally reside in the intestinal tract. In the past, *Salmonella* infections were usually traced to dirty or cracked eggs contaminated from the outside with chicken droppings; however, *Salmonella* organisms can now be found inside intact eggshells. In 1991, *Diseases of Poultry* observed: "With the great expansion of the poultry industry, the widespread occurrence of avian salmonellosis has ranked it as one of the most important egg-borne bacterial diseases of poultry, [which] constitutes the largest single reservoir of *salmonella* organisms existing in nature" (Calnek, 72).

In addition to crowding and filth, the U.S. egg industry's practice of withdrawing food from hens for days and weeks at a time, known as forced molting, was shown in the 1990s to facilitate *Salmonella enteritidis* infection of hens' oviducts and their eggs—an infection that can be transmitted to humans.

It appeared that the hens' immune systems couldn't cope with the multitude of stressors, including *Salmonella*-infested rodent droppings in their food. "One rodent can deposit 100 pellets in the course of one night and each pellet can contain 25,000 different salmonella organisms," a United Egg Producers spokesman told an FDA-USDA Egg Safety Meeting in Columbus, Ohio, in 2000 (67). So severe was the *Salmonella* infection in force-molted hens that, according to researchers, "Molting, in combination with an SE infection, created an actual disease state in the alimentary tract of affected hens whereas, under normal conditions, little SE-induced morbidity occurred in adult birds" (Holt and Porter, 1842).

In 1993, United Poultry Concerns launched a campaign to educate the public about forced molting as both an animal cruelty and a *Salmonella* food poisoning issue. Working with the Association of Veterinarians for Animal Rights, we sought to get the U.S. egg industry to stop starving hens and to get the American Veterinary Medical Association to stop endorsing the practice, which they did in 2004 ("Delegates act," 5–6).

In 2005, our efforts succeeded in "reversing a practice followed for more than 100 years," according to *Feedstuffs* (Smith 2005, 1). From then on, United Egg Producers would encourage its members to restrict hens' food instead of taking it all away. Researchers reported in 2006 that food restriction rather than food deprivation appeared to reduce the risk of "SE contaminated eggs entering the human food supply" (Webster 2006).

Salmonella-infected hens and eggs is a continuing problem since dead birds and manure are continuously recycled into products for feed and fertilizer (Kiepper). As noted in *Feedstuffs*, "Despite the many potential sources of contamination, poultry feed, especially that containing animal by-products, has long been incriminated as a primary source of salmonella contamination" (Jones and Richardson).

In 2004, the *Wall Street Journal* reported that an estimated 118,000 people in the U.S. become sick each year "from ingesting eggs contaminated with salmonella bacteria, which can cause mild to severe gastrointestinal illness, short-term or chronic arthritis and, in rare instances, death" (Zamiska). The U.S. Centers for Disease Control and Prevention (CDC) states that "for any culture-confirmed case of salmonella or any pathogen that's reported to CDC, many more cases go unreported." Thus a single confirmed case of SE in the entire country "actually represents 38 cases in the general population" (FDA-USDA, 4).

The problem of *Salmonella*-contaminated eggs goes beyond the United States. A 2005 study in the European Union on commercial laying hen operations with at least 1,000 hens per flock found *Salmonella* in one-third of the operations. The most prevalent was *Salmonella enteritidis* followed by four other types of *Salmonella* bacteria that can make people sick. According to the study, "Salmonellosis along with campylobacteriosis, are by far the most frequently reported food borne diseases in the EU," primarily "through ingesting poultry and poultry products, such as eggs" ("Study finds Salmonella").

Salmonella and Antibiotics

Antibiotics are used to control the bacterial diseases that thrive in crowded confinement and to manipulate egg production. For example, Neoterramycin is given to laying flocks suffering from infectious syndromes such as peritonitis in the abdominal cavity, airsacculitis in the respiratory organs, and inflammatory salpingitis in the oviducts. *Salmonella*, *E. coli*, *Staphyloccus*, and other bacteria are involved in these syndromes affecting caged laying hens (Shane 2006a).

The link between *Salmonella*, antibiotics, and global concentrations of birds was described at the International Egg Commission's 1995 conference in Stockholm, Sweden:

> [A]t the end of 1980 a dramatic change occurred when a certain serotype of salmonella (*S. enteritidis*) changed its properties so it could infect the egg and/or cause disease in humans with relatively few bacteria. Due to the global concentration of the production of breeding birds, this change resulted in a severe worldwide epidemic of Salmonellosis in humans, demonstrating the necessity to control salmonella in layers producing eggs for human consumption. . . . In addition, the often uncontrolled use of antibiotics, as an easy way to control the situation, in combination with the great ability of the more than 2,000 different types of salmonella bacteria to adapt to new situations, have contributed to the current serious situation. Of the food-producing animals, poultry is most often considered as the main source (Evans 1995).

With increased concern over the role of antibiotics in creating bacterial resistance to medical treatment, McDonald's announced a ban on the use of growth-promoting antibiotics in chickens raised for the company by the end of 2004, and the European Union is considering a ban on antibiotics on poultry farms by 2012 (Davis 2003a). In 2005, the U.S. Food and Drug Administration banned the use of the fluoroquinolone antibiotic Baytril in poultry flocks after having approved its sale in 1996 to treat respiratory diseases in factory-farmed chickens and turkeys. According to *The Washington Post,* this was the first time the FDA succeeded in "forcing off the market an antibiotic used to treat animals because of concerns that it will make similar antibiotics less effective in treating people." The Union of Concerned Scientists estimated that "24.6 million pounds of antibiotics are used on American farms yearly, about 75 percent of the nation's antibiotic consumption" (Kaufman 2005).

Manure and Ammonia Burn in Pullet and Layer Houses

Manure is by far the number one waste problem.

—Bell and Weaver, 149

Ammonia gas is the most common eye irritant that domestic birds encounter. Ammonia gas is inevitably produced in the poultry house when uric acid from bird waste combines with water in the environment. . . . Ammonia gas is extremely irritating to the membranes that line the eyelids, eyes, sinuses, and trachea.

—Patricia Dunn, DVM

Manure is everywhere in the caged layer complex. Toxic ammonia rises from the decomposing uric acid in the manure pits beneath the cages to produce a painful corneal ulcer condition in chickens known as ammonia burn, a keratoconjunctivitis, which can lead to blindness (Calnek, 852). Affected birds have "reddened, swollen eyelids. They appear to have pain around the eye and attempt to avoid light. If high ammonia levels persist, the cornea, the outermost part of the front of the eyeball, becomes ulcerated and blindness occurs" (Dunn).

Ammonia facilitates chronic respiratory diseases in chickens including Newcastle disease, laryngotracheitis, infectious bronchitis, Bordetella, Haemophilus, Chlamydia, Mycoplasma, and Cryptosporidium (Carlile; Dunn). Ammonia fumes injure the mucous membranes of the upper respiratory tract making it easy for bacteria, viruses, and protozoa to invade and colonize the lungs, air sacs, and livers of exposed birds. Ammonia fumes entering the blood cause immunosuppression, which further encourages disease organisms to colonize and spread through body parts. Studies suggest that even at low concentrations, significant quantities of ammonia can be absorbed into an egg (Carlile, 101).

The hens spend their lives in the presence of ammonia. The 220 million hens hatched for table-egg production in the United States each year occupy pullet cages, layer cages, and transport cages. According to *Commercial Chicken Meat and Egg Production*, "As early as the 1930's, young chickens were kept in batteries, cages, and wire-floored pens on commercial farms. Today, practically all new egg-type pullet-growing facilities built in the United States are of the cage type" (Bell and Weaver, 20; 979-1006).

The baby chicks grow to egg-laying maturity in cages stacked from one to four decks high. Many of these future battery-caged hens are raised in dark-out houses as well, where no natural sunlight is allowed to enter, in order to control their sexual development, because daylight stimulates egg production. As in the battery-cage houses to which they will be moved at 18 weeks old, droppings "fall through the mesh bottom of the cage to the house floor below" and may drop directly through the lower cages onto the birds in those cages (Bell and Weaver, 979–1000).

According to a researcher, a one-million-hen complex produces 125 tons of wet manure a day. For every truckload of feed brought into the farm, "a similar load of waste must be removed" (Bell 1990). For every 700,000 hens, 1,500 birds die each week in their cages: "At a three-pound average, that's more than two tons of dead chickens to haul off each week—well over 100 tons a year, at more than $12,000 in transportation and landfill costs," a University of Georgia report stated ("Environment").

"Broiler" chickens—birds used for meat as opposed to egg production—are raised on floors and slaughtered as babies; thus broiler sheds could be cleaned out every six weeks, though in fact they are not. However, laying hens are confined in the same building for one or two years in stacked cages, which raises the question of how to remove the manure and the corpses without disturbing production. Even when a house is temporarily empty, as a Michael Foods employee told an egg safety meeting in Columbus, Ohio, "to wash a house takes at least two weeks, eight to ten people, and nearly 24 hours a day washing per day to get it clean" (FDA-USDA, 19).

Despite the fact that manure fumes and rotting carcasses force workers in the battery houses to wear gas masks, the egg industry claims that the cage system is more hygienic than free-range and floor systems, because hens in cages have less direct contact with their own droppings, which are (in principle in some operations) deflected by a device to the pits beneath the tiers of cages. A researcher told the agribusiness newspaper *Feedstuffs*, in 2006, that hens in cages "can't flap their wings, but they are protected from the dust [and] poor air quality" of non-cage systems (Smith 2006). This is so absurd. Anyone who has ever been inside a caged laying hen facility can testify to the stale, stinking, nauseating eye and throat burning stench, the feathers, dust, and dander floating and suspended in the air, the mountain of manure beneath the cages, the hens huddled with feces on their feathers and raw skin, the cage bars encrusted and dripping with droppings from the cages above, the cobwebs, flies, and rodents.

The egg industry does not want to give up cages, claiming dust and manure buildup among reasons. An article in *Feedstuffs* cited a visitor to a cage-free facility (in which hens are confined in buildings but not in cages) who said "there was so much dust in the house that he could barely see the chickens" (Smith 2006). Why? When birds are crowded together, filth accumulates on the floor and the ground, and in the air and the water, and is passed around by the simple act of breathing.

Organisms like coccidia and *Salmonellae* thrive in the dampness that develops. Chickens do not choose to be dirty. Given a chance, they regularly dustbathe and preen to keep their skin fresh and their feathers soft and lustrous.

But even if the chickens can go outside, the floor of a crowded hen house and the ground surrounding it will become contaminated—"fowl sick." The more chickens there are in less space, the more manure there will be. Soon there is more manure and microbes than the litter inside and the land outside can healthfully accommodate. The air becomes saturated with ammonia, flies buzz, and the place is a mess.

In contrast, fewer chickens with more space means that fewer droppings will be dispersed over a wider area to be dried by the sun and absorbed into the soil. The mineral-rich droppings of a small chicken flock benefit the land. The scratching of the earth by healthy, ranging fowl improves soil fertility. In the film *The Bird Man of Alcatraz,* based on a true story, diseases emerged only after the "bird man" began crowding his prison cell with birds. The cures he invented were for diseases he promoted. This is the type of "progress" of intensive poultry and egg production.

Coccidiosis

In spite of many excellent programs and chemicals for the control of coccidiosis, there are many outbreaks of the disease.

—Bell and Weaver, 483

Take, for example, coccidiosis. This disease is caused by coccidia, a protozoan parasite, which, under normal circumstances, lives harmlessly in the gut of chickens and other birds and is shed in their droppings. Birds become exposed to coccidiosis by picking up the sporulated oocysts in these droppings.

Historically, coccidiosis was not a problem. A drug company explains: "Back around 5400 B.C., when the first domestic chickens appeared, coccidia were right there with them. And nature maintained a healthy balance between the two species. Until modern man upset it by raising birds in confinement" (Elanco).

An article in *Poultry World* states: "Coccidiosis is typically a disease linked to intensive animal production. The reason for outbreaks is that we stock a high number of young, susceptible animals in an environment which is ideal for the reproduction of the coccidia" (van der Sluis 1993, 16).

A chicken breeder from the 1920s recalls that coccidiosis "was just becoming a factor, because of the large number of birds being grown in one place" (Coleman 1976, 51).

Today, coccidiosis is an infectious disease "endemic to the poultry industry." Its symptoms include "poor appetite, listlessness, lowered feed efficiency, and diarrhea, potentially leading to mortality. The parasite invades the intestinal wall, causing lesions that impair nutrient absorption" ("USDA approves"). There are many coccidiosis drugs known as coccidiostats, or anticoccidials, on the market. One, known as lasalocid, was described in 2002 as a "potentially dangerous drug present in up to 750,000 eggs eaten in Britain every day [that] can cause heart contractions in humans and builds up in the body every time we eat eggs or chicken" ("Campaigners say"). In 2005, the Soil Association published a report, "Too hard to crack—eggs with residues," explaining that lasalocid, which is sold worldwide as a poultry feed additive known as Avatec 15% by the pharmaceutical company Alpharma, is not legally permitted in laying-hen feed in Britain, "but it is permitted for young birds destined to become layers." Lasalocid accumulates in egg yolk and has been found in free-range, barn-laid, and battery-cage eggs, according to the Soil Association (2–3, 6).

Coccidiosis exemplifies the disease-producing nature of the modern poultry and egg industries. An article in the 1970s observed that as the poultry industry grew, so did "the incidence of poultry disease. Greater concentrations of birds within a house, larger houses with more birds, and a greater bird density within a given land area was causing high mortality" (North, 88).

"Cannibalism"

Feral chickens spend about half their time foraging and feeding, and make an estimated 14,000–15,000 pecks at food items and other objects in the course of a day.

—Bell and Weaver, 73

Cannibalism develops either as a result of misdirected ground pecking or is associated with dustbathing behaviour.

—Philip C. Glatz, *Beak Trimming,* 11

In cages, feather pecking occurs particularly during the afternoon when hens have finished feeding and laying eggs and have little else to do.

—Christine Nicol and Marian Stamp Dawkins,
"Homes Fit for Hens," *New Scientist*, 50

North and Bell state, "When birds are given limited space, as in cages, there is a tendency for many to become cannibalistic" (309). This behavior is caused by the abnormal restriction of the normal span of activities in ground-ranging species of birds in situations in which they are squeezed together and prevented from exercising their natural exploratory, food-gathering, dustbathing, and social impulses. It includes vent picking, feather pulling, toe picking, head picking, and in some cases consumption of flesh. Pecking in chickens is a genetic behavior developed through evolution that enables them not only to defend themselves, but, as Rogers states, "birds must use the beak to explore the environment, much as we use our hands" (1995, 96).

Poultry researchers attribute abnormal pecking behavior to a variety of interactive causes, including artificial lighting intensities and durations to force overproduction of eggs. In addition, chickens may be driven to peck and pull the feathers of cagemates in order to obtain nutrients they would find when ranging but cannot obtain in fixed commercial rations. Mash and pellets exacerbate the problem because the birds cannot select specific nutritional components; they fill up quickly and are left with nothing to occupy their beaks and their time (Rogers 1995, 219; Bell and Weaver, 86, 239).

Caged chickens are also driven to peck at each other as a result of their inability to dustbathe. Studies show that hens deprived of dustbathing material suffer from "an abnormal development of the perceptual mechanism responsible for the detection of dust for dustbathing." Without any form of loose, earthlike material, chickens "are more likely to come to accept feathers as dust" (Vestergaard 1993, 1127, 1138).

In addition, research shows that fear is not only a result of the pecking of cagemates, but also a cause of it. According to Klaus Vestergaard, "the peckers are the fearful birds, and the more they peck the more fearful they are. This finding emphasizes abnormal behavior in the evaluation of well-being in animals which have no obvious physical signs of suffering" (1993, 1138).

A poultry researcher learned the importance of pecking when he designed a method of feeding chicks by pumping slurry directly into their necks. He wrote, "The slurry that was fed had the right amount of both food and water, so that the chicks did not need to peck to prehend either feed or water. And, the end of all that research (with Dr. Graham

Sterritt, an NIH Fellow) for 5–6 years was that chicks peck independently of whether or not they need to peck in order to eat. I still cannot believe all of the money spent to study that" (Kienholz 1990).

A poultry breeder recalls the emergence of cannibalism in the 1920s. "When I was a senior, the University [of New Hampshire] hired a new laboratory man from the West—Dr. Gildow. He recommended using wire platforms, which let the droppings fall through the floor. This interrupted multiplication of the [coccidia] oocysts. But it led to a completely new problem—cannibalism—and after a year or so wire platforms were out" (Coleman 1976, 51). Out, that is, for broiler chickens, but not for laying hens.

A poultry nutritionist recalls that when high energy feeds "were not adequately fortified with other nutrients, especially protein, they caused a new problem—cannibalism and feather picking. The problem was aggravated by excessive use of supplementary light. . . . Debeaking helped to control cannibalism and soon became a standard practice" (Day, 142).

DEBEAKING

The emotion-laden word "mutilation" is sometimes used in describing husbandry practices such as removing a portion of a hen's beak. . . [However] removal of certain bodily structures, although causing temporary pain to individuals, can be of much benefit to the welfare of the group.

—Craig, 243–244

Feather pecking is not aggression; rather it's foraging behaviour gone wrong. The solution of industry is to chop off beaks.

—Ian Duncan 2006

The trigeminal ganglia, the site of the first order of sensory neurons that innervate the face and beak, develop when the embryo is two days old.

—Glatz, 47

The integument of the chicken (skin and accessory structures, e.g., the beak) contain many sensory receptors of several types allowing perception of touch (both moving stimuli and pressure stimuli), cold, heat, and noxious (painful or unpleasant) stimulation. The beak has concentrations of touch receptors forming specialized beak tip organs which give the bird sensitivity for manipulation and assessment of objects. Beak trimming deprives the bird of normal sensory evaluation of objects when using the beak.

—Bell and Weaver, 80

DBK 2000 Precision Beak Trimming
• trim 2 birds at a time
• up to 2,000 birds per hour per two man team
• accurately control cauterization time and temperature
• accurately control chick count per cage
• ability to remove cage liners at time of beak trimming
• ability to separate chicks with accurate chick count at time of beak trimming
• operator comfort air ride seat
From Baer Systems, Inc., Lake Park, MN 56554
Advertised in *Egg Industry*, January 2008

Egg producers remove up to two-thirds or more of hens' beaks with a hot or cold machine blade, without painkillers, to reduce "cannibalistic" pecking and lower the cost of feeding the birds. Debeaked birds have been shown to suffer acute and chronic pain and distress (Gentle 1990; Duncan 1993). Their appetites are reduced, and they do not grasp their food efficiently, which causes them to eat less, fling their food less, and "waste" less energy than intact birds, thereby (it is claimed) saving the industry money (Craig 1992, 1830). Rough handling by operators, including making noise, yelling, and grabbing the birds by the head, neck, wing, or tail as they shove the birds' faces up against, or into, the debeaking machinery, then pull the birds violently away and toss them into containers, causes broken bones, torn and twisted beaks, hidden joint damage, and other injuries (Glatz, 87–92; Ruszler, 2).

Debeaking—also referred to as partial beak amputation or beak trimming—began in the 1930s and '40s when a "gas torch was used by T. E. Wolfe in San Diego County in California to burn off part of the upper beak of the hen." Later, a neighbor of Wolfe, W. K. Hopper, adapted a "tinner's soldering iron by giving it a chisel edge, which enabled the operator to apply downward pressure on the upper beak to sear and cauterize the beak." The Lyon Electric Company in San Diego adopted some of these modifications to develop the first beak-trimming machine. The company "first brought out a heated knife attachment for a homemade beak support and frame. The name for the machine, 'debeaker,' was coined in 1942 and registered in 1943" (Glatz, 3).

Chickens raised for meat are no longer debeaked because "meat-type" chickens are slaughtered as six-week-old babies, before they are old enough to form a social order. North and Bell explain, "Because social dominance in either sex is not evident prior to 8 to 10 weeks of age, it is possible to raise large numbers of broilers in a single pen without fear that the birds will develop social agonistic characteristics" (544–545).

In contrast, hens used to produce eggs for human consumption, broiler breeder roosters, and egg-industry roosters and hens used for breeding are debeaked, one or more times, by contract or company teams, between the ages of one day old and five months old, before egg laying begins. (Likewise, turkeys, pheasants, quails, and guinea fowl are debeaked and ducks are debilled.) Because a severed young beak can grow back or be fatally injured, "the most popular age for beak-trimming," according to Glatz, is "from 5 to 10 days of age," even though "research indicates that day-old beak-trimming causes the least stress" (2, 87).

Lyon Electric Company warns customers that failure to debeak "properly" can cause "starve-outs," feed wastage, and even the so-called cannibalism that debeaking is supposed to prevent. Poultry manuals tell farmers that if an electric beak trimmer isn't handy, "a sharp jackknife or a pair of scissors" may be temporarily used. Researchers in the 1990s experimented with jackknives, dog nail clippers, and pruning shears ("secateurs"). For example, Grigor, Hughes, and Gentle (1995) "used a pair of secateurs at 1, 6, or 21 days to trim the upper beak of turkeys. There was bleeding from the upper mandible, which ceased shortly after the operation. Despite the beak regrowth a reduction of cannibalism was noted," says Glatz, without explanation (5). In another experiment, Gentle, Hughes, Fox, and Waddington (1997) "used secateurs to remove one-third of the upper beak in Isa Brown chickens." Summarizing many debeaking experiments using blades, Glatz says there were "very few differences observed between behaviour and production of the hot blade and cold blade trimmed chickens" (Glatz, 5–6).

The poultry industry used to deceive the public that a chicken's beak was as insensitive as the tip of a human fingernail, but this assertion can no longer be made because decades of research have refuted it. In fact, debeaking was fully explored by the Brambell Committee, a group of veterinarians and other experts appointed by the British Parliament to investigate animal welfare concerns arising from Ruth Harrison's exposure of factory farming in her book *Animal Machines*, published in 1964. Reporting on farmed animal welfare in the U.K in 1965, the Brambell Committee recommended that "beak-trimming should be stopped immediately in caged birds and within two years for non-caged birds" (Glatz, v).

The Committee stated that irrespective of whether the operation is performed competently, and in the way that meets with the general approval of the poultry industry, debeaking is not similar to fingernail clipping: "The upper mandible of the bird consists of a thin layer of horn covering a bony structure of the same profile which extends to within a millimeter or so of the tip of the beak. Between the horn and bone [of the beak] is a thin layer of highly sensitive soft tissue, resembling the quick of the human nail. The hot knife blade used in debeaking cuts through this complex of horn, bone, and sensitive tissue causing severe pain" (Mason and Singer 1990, 39–40).

"Acute and Chronic Pain"

In 1992, Ian Duncan, a poultry researcher at the University of Guelph in Ontario, explained why "there is now good morphological, neurophysiological, and behavioral evidence that beak trimming leads to both acute and chronic pain."

> The morphological evidence is that the tip of the beak is richly innervated and has nociceptors, or pain receptors. This means that cutting and heating the beak will lead to acute pain. In addition, it has been shown that as the nerve fibers in the amputated stump of the beak start to regenerate into the damaged tissue, neuromas form. Neuromas are tiny tangled nerve masses that have been implicated in phantom limb pain (a type of chronic pain) in human beings.
>
> The neurophysiological evidence is that there are abnormal afferent nerve discharges in fibers running from the amputated stump for many weeks after beak trimming—long after the healing process has occurred. This is similar to what happens in human amputees who suffer from phantom limb pain.
>
> The behavioral evidence is that the behavior of beak-trimmed birds is radically altered for many weeks compared to that which occurs immediately before the operation and compared to that shown by sham-operated control birds. In particular, classes of behavior involving the beak, namely feeding, drinking, preening, and pecking at the environment, occur much less frequently, and two behavior patterns, standing idle and dozing, occur much more frequently. The only reasonable explanation of these changes is that the birds are suffering from chronic pain (Duncan 1993, 5).

Based on the demonstrations of neuropathic pain, sickness behavior, and other criteria of suffering and debilitation in beak-trimmed birds, Farm Animal Welfare Council, a government advisory organization in Britain, reaffirmed the Brambell Committee's findings from the 1960s in its 1991 report *Welfare of Laying Hens in Colony Systems*, stating that debeaking is "a serious welfare insult [injury, attack, or trauma] to the hens" that "should not be necessary in a well-managed system where the hens' requirements are fully met" (FAWC, 23–24). However, debeaking is still being done in the European Union, with the exception of Sweden, where the procedure is banned. A 2004 report by Compassion in World Farming Trust titled *Practical Alternatives to Battery Cages for Laying Hens* lists the "banning of beak trimming that causes both acute and chronic pain" as a major legislative objective not yet obtained (CIWF Trust 2004, 2). Currently, there is no law or pending legislation prohibiting debeaking in North America. There needs to be.

Poultry producers know that debeaking causes pain. They have their own term, "beak tenderness," to describe the condition that prompts advice about such things as the need for deep feed troughs to prevent the wounded beak from bumping the bottom of the trough, resulting in starve-outs: "Striking the tender beak would certainly be a deterrent to normal feed consumption" (North and Bell, 250). Debeaking machine operators are reminded to do the "very tedious task" of beak trimming carefully. "Too often it is done carelessly. . . . Be sure not to sear the eyes when trimming" (North and Bell, 246, 248, 251). And remember: "An excessively hot blade causes blisters in the mouth. A cold or dull blade may cause the development of a fleshy, bulb-like growth on the end of the mandible. Such growths are very sensitive and will cause below average performance" (Thornberry, 207).

Debeaking does not stop "cannibalism" anyway. *Diseases of Poultry* states that a "different form of cannibalism is now being observed in beak-trimmed birds kept in cages. The area about the eyes is black and blue with subcutaneous hemorrhage, wattles are dark and swollen with extravasated blood, and earlobes are black and necrotic" (Calnek, 827). Glatz says improperly cauterized birds who bleed after beak trimming may attract other birds "to gently peck at the wound," encouraging them later to "initiate cannibalism" (Glatz, 89).

"Further Research"

Beak-trimming experiments are a worldwide enterprise. *Beak Trimming* contains 27 pages of published experiments spanning four decades. The Brambell Committee report, published in 1965, urged that beak trimming be immediately discontinued on welfare grounds, only to be followed by countless more experiments (many unpublished) continuing into the twenty-first century. Despite the "wealth of scientific information on the welfare of beak-trimmed birds, beak-trimming methods and alternatives to beak-trimming" (Glatz, v), researchers insist "there is a lack of comprehensive studies that measure the effect of beak-trimming on welfare using multiple indicators (physiological as well as behavioural) and it is hard to compare between studies due to different methods of beak-trimming and beak-trimming at different ages" (Glatz, 77).

These different methods include—in addition to the hot blades, cold blades, soldering irons, jackknives, pruning shears, and dog nail clippers already mentioned—liquid nitrogen used to "declaw emus"; gas beak-trimming machines consisting of "a hot plate and cutting bar operated by means of a foot lever" and available as a "pocket-style machine for trimming pullets which uses gas from a cigarette lighter as its heat source"; robotic beak trimmers, where chicks are "loaded onto the robot by hand being held by cups around their heads"; chemical beak-trimming using capsaicin, "a cheap nontoxic substance extracted from hot peppers [that causes] depletion of certain neuropeptides from sensory nerves in birds"; infrared beak treatment machines that cause the affected part of the beak to soften and "erode away" within a week; and laser machines, which operate "by sending energy to the target tissue," cutting the tissue (often unsuccessfully) with "intense emissions of light" and heat absorption.

That these procedures are horribly painful can be seen from the cold-hearted descriptions. For example, day-old chicks subjected to an ophthalmic laser "vocalized" in response to an increase in "energy density," indicating they were feeling more "discomfort" than chicks used in another laser surgery to which this experiment is compared. The "discomfort" is thus explained: "The laser was able to cut through the outer layers of keratin, but was not able to cut the inner bone. The lack of success in being able to cut the bony portion of the beak with an ophthalmic laser was considered to be due to the lack of [electrical] power." "Further work is now required . . ."(Glatz, 9).

There is also the Bio-Beaker. Developed in Millsboro, Delaware, in the 1980s, it uses a high-voltage electrical current to burn a small hole in the upper beak of chickens and can beak-trim up to 2,000 day-old chicks in an hour. According to Glatz, "The chicks being bio-beaked struggle as the beak is inserted into the mask of the instrument and also when the current is passed." Notwithstanding, the fact that the Bio-Beaker achieves "an adequate beak-trim during the first day of life [makes] the unit ideal for use in the hatchery. Unfortunately in many chicks the tip of the beak did not slough off and birds had to be re-trimmed using conventional equipment."

Used on turkeys, the Bio-Beaker is said to be "more successful with the beak tip falling off in 5–7 days and the wound healed by 3 weeks." The Bio-Beaker is used for trimming the upper beaks of commercially raised turkeys, even though "operator errors and inconsistencies have caused welfare problems for turkeys" (Glatz, 10). Perhaps the Bio-Beaker is responsible for the blackened, necrotic, crumbling beaks of baby turkeys photographed by investigators in recent visits to some U.S. turkey farms ("Turkey Beak Image").

White Strings and Peck-O-Meters

Research is being done to see if "cannibalistic" pecking can be prevented without debeaking and without sacrificing productivity and profits. Proposed alternatives involve environmental enrichment and genetic manipulation. Environmental enrichment refers to confinement furniture—nest boxes, perches, and dustbathing areas and/or "stationary bunches of plain white string" for birds to "tease apart in a way that resembles preening"! One wonders how long these strings would stay white in the filthy houses where white-feathered hens quickly become sepia-colored, or with what eagerness the egg industry would "consider developing an automated management system that was capable of detecting waning of interest in [string] enrichment devices and then raising them briefly or moving them to a nearby location in an attempt to rekindle interest" (Glatz, 99).

As for genetic selection against "cannibalism," here's an example:

> Indirect selection for and against feather pecking was made at the University of Hohenheim, Germany. Strains of laying hens were selected for or against, pecking at a bunch of feathers connected to an automated recording system (Figure 8.2.2), the 'peck-o-meter' (Bessei, Reiter, Bley, and Zeep 1999). The preliminary estimates of heritability in generation 1, 2, and 3 respectively were 0.19, 0.22, and 0.26 (W. Bessei, K. Reiter and A. Harlander-Matauschek, pers. com.). With regard to feather pecking (on live animals), the line selected for a high level of pecking at the feathers of the "peck-o-meter" showed less feather pecking compared to the line selected for a low level of

> pecking. The negative phenotypic correlation, estimated to be about -0.30, between pecking at the "peck-o-meter" and feather pecking is the opposite of the expected relationship according to correlations obtained by (Bessei et al., 1999) and needs further investigation. Other studies have also reported lack of correlation between feather pecking and other behaviour patterns that might have been used for indirect selection (Albentosa, Kjaer and Nicol 2003; P. Locking, pers. com.), so at the moment only direct selection can be recommended (Glatz, 105).

The egg industry says that even if breeders breed more placid hens, "beak trimming may still need to be considered for economic reasons for the reduction of appetite and feed wastage" ("Feed Savings"). Poultry welfare scientist Ian Duncan told a conference in 2006 that while selective breeding could eliminate feather pecking, breeding companies aren't interested because they might have to sacrifice production.

Dustbathing

Galliformes, the bird family which includes domestic hens, bathe only in dust, unlike other types of birds which bathe in both water and dust. . . . Depriving hens of access to litter results in the development of sham, or vacuum, dust bathing, in which the hen goes through the motions of dust bathing on a bare floor. Dust bathing motivation gradually builds up over the deprivation period in the afternoon, their natural dust bathing period.

—Michael R. Baxter, 616

An increase in the amount of pre-dustbathing vocalizations was found after the long deprivation [7 days], particularly in the White Leghorn hens.

—Koene and Wiepkema

It is unethical to deprive an individual creature the opportunity to practice bodily hygiene.

—Karen Davis, *Egg Industry*, October 2002, 17

An example of a natural behavioral need of chickens that is cruelly frustrated by cage confinement is dustbathing. Chickens, turkeys, and other ground-nesting birds dustbathe to clean and refresh themselves, distributing loose earth and the oil from the preen gland at the base of their tail through their feathers and onto their skin to remove built-up oil, dead skin, and skin irritants, and to maintain and improve feather structure (Baxter, 616; Vestergaard 1981, 7). Chickens and turkeys dustbathe frequently in hot weather in order to cool themselves and during the autumn molting of their feathers, which "appears to produce an itching in the skin" to which frequent dustbathing gives relief (Schorger, 177).

Chickens released from a cage to a suitable area will immediately start making a dust bowl, paddling and flinging the dirt with their claws, raking in particles of earth with their beaks, fluffing up their feathers, rolling on their sides, pausing a while from time to time, and stretching out their legs in obvious relish. Vestergaard states: "Dustbathing is another example of a behavior which has been ignored—perhaps not even known—by designers of battery cages."

> Dustbathing is a complicated behavior which takes about half an hour. It is a significant part of feather maintenance of gallinaceous [ground-nesting] birds. The behavior occurs in all housing systems for hens, including battery cages, even where there is no "dust" (sand or litter). In these cases, dustbathing is often distorted, since the hen tries to direct the bill-raking pattern toward the food, while taking a posture between half standing and half sitting. The violent movements toward the wire floor may cause pain and damage to the feathers, and the other birds often peck violently at a dustbathing bird. Experiments have shown that during deprivation from sand, dustbathing motivation increases. . . . So, again, we are here dealing with behavior and a motivation system which has evolved in nature but cannot be manifest, and results in trouble for the animal (Vestergaard 1987, 10).

In addition, as Ruth Harrison explained in an article in *New Scientist*, even the type of dustbathing material provided for hens in a noncage confinement environment affects their well-being.

> [T]he type of litter and depth of litter, as well as quantity of litter, have a vital part to play in reducing feather pecking and encouraging adequate scratching and dustbathing behaviour, and even in keeping the birds warm in winter. Wood shavings are the most commonly used litter in alternative systems today, but they are also the least satisfactory because they adhere to the outer

feathers and do not penetrate to the skin to assist in removal of excess oil and ensure that the plumage remains in good condition. Recommended alternatives are peat, chopped straw, and sand. . . . Recent work shows that hens in bright light (500 lux) will work litter far more than hens in low light (50 lux), keeping it friable rather than compacted and damp—conditions that allow a build-up of microorganisms in the litter, making the birds more vulnerable to disease (Harrison 1991, 43).

Heat Stress

Birds in cages cannot withstand hot climactic conditions as well as birds on a litter floor; caged hens are completely surrounded by hot air, and have no way of getting away from the heat.

—Bell and Weaver, 1026

The ultimate trouble for the caged hen is that she is forced to live in a world that makes no sense to her nature. She did not choose it, she cannot escape it, and she cannot change it. The caged environment reflects human psychic patterns, not hers.

Chickens dustbathe not only to clean but to cool themselves. They do not perspire, so on hot summer days they pant with their beaks open, hold their wings away from their bodies, and dustbathe beneath a shady tree or other refreshing cover. When the temperature reaches 80 degrees F (27 degrees C), chickens start to suffer. They develop heat stress—physiological responses to remove excess deep body heat. If the situation continues, they lose immunity because the bursal cells responsible for immunological competence are heat sensitive. The result is something like AIDS in humans (Coleman 1995).

The main sources of heat in a caged layer house are the hen's own body heat, multiplied many thousands of times, and the manure mounds beneath the cages (Duncan 2006). When the house gets hot the hens cannot properly rid themselves of this heat. Body heat mounts. When it reaches about 117 degrees F (47 degrees C), chickens die (Muirhead 1993a). Fans, foggers, roof sprinklers, and evaporator pads are installed to reduce the high mortality rate and drop in egg production that occur in hot weather, However, poultry units are not air-conditioned; they have nipple drinkers instead of troughs, so the birds can never drink deeply, and when the fans and foggers break down, the birds are stuck. In addition, the high levels of chloride bleach in the water used on poultry farms has been shown to reduce birds' water consumption, resulting in higher heat stress mortality (Ernst).

Every summer, millions of hens die of heat stress (and in fires) trapped inside their cages. In the summer of 1995, an estimated three to five million hens died in the heat wave that spread through the eastern half of the United States ("Heat Deaths"). In July of 2006, more than 35,000 hens died in their cages when Rose Acre Farms in Newton County, Indiana, cut off the airflow to fight a fire in another building (Wyatt). Even when the electrical systems work, the stocking density turns the houses into ovens. Industry literature advises less crowding as a way to reduce the summer heat load, but adds reflexively that this "may not be economically feasible under many commercial conditions" (Kreager, 33).

Mash, Mold Toxins, and Mouth Ulcers

Hens suffering from heat stress stop eating, but eating is difficult regardless because of debeaking and because the battery-caged hen must stretch her neck across a feeder fence to reach the monotonous mash in the trough, a repeated action that over time wears away her neck feathers and causes throat blisters. In addition, the fine mash particles stick to the inside of her mouth, attracting bacteria and causing painful mouth ulcers (Gentle 1986).

Adult chickens require food particles of varying sizes and shapes for oral hygiene. They "prefer the feel of large particles in their beaks." However, *Feedstuffs* explains: "Hens fed coarse meals devour profits. . . . [T]here is excessive food 'usage' without any improvement in laying performance" (Phelps). As if all this were not enough, certain mold toxins called mycotoxins (aflatoxins and T-2 toxins) poison the mash in hot, humid weather causing the hens to develop mycotoxicosis, or fungal poisoning. Egg production drops and immunity is suppressed (Whitlow and Hagler, 77). Hens develop mouth ulcers; pale facial appearance; high disease susceptibility; hemorrhaging of kidneys, lungs and heart; bruising; and bloody thighs (Behrends).

Forced Molting

Replacements represent the second largest cost of producing eggs; feed is first... It costs less to molt a flock and bring it back into egg production than it does to grow a new flock.

—Bell and Weaver, 1059, 1062

Our chicken houses hold 126,000 give or take a few hundred. Our molts usually last about 12 days and during the molt we lose right around 50 birds a day. The average mortality per day is around 20 or 30. The last couple of days of the molt before we feed them we lose 100 to 150. The day we feed them we will lose about 200–250 hens within a few hours after we feed them. The hens tend to gorge themselves and choke on the feed as they try to eat too much too soon, or at least that's what we believe. The following day we usually plan on losing 500 to 1,000 depending on how well we carry out the molt. The days following, our mortality is usually cut in half each day. . . . The ammonia in the house during this period is so bad that we usually wear masks in order to breathe. The ventilation in the building is minimal during a molt. In fact, the ammonia and dust during a molt is just as bad in the buildings as it is during the winter in all the buildings. It is almost unbearable to us. I imagine it is just as bad for the hens.

—Chris Hernandez, Cal-Maine equipment manager, 1999

We passed on through the egg barn. When the lights came on, the cackling and clucking rose to a cacophony, accompanied by the sound of thousands of beaks pecking on metal.

—Kathy Geist

The U.S. poultry and egg industries use food deprivation and nutrient restriction as an economic tool to manipulate egg production in commercial laying hens and in male and female birds used for breeding of both egg-type and meat-type birds (Bell and Weaver, 1059–1077). In the United States, hens used for commercial egg production are depopulated at seventeen or eighteen months old, or they are kept for another laying cycle and depopulated at two years old. The dwindling number of survivors may even be kept for a third cycle until they are two and a half years old, and then destroyed, whichever is cheaper (Bell and Weaver, 1061).

Birds to be reused are force-molted—"recycled"—to prepare them for the next laying cycle. In this procedure, they are partially or completely starved for two to five to fourteen days or longer to give them a "rest" (Duncan 1999; Bell 2005). Their food is removed or nutritionally reduced, causing the hormone levels that induce egg production and inhibit feather growth to drop. New feathers push out old ones, and the hen stops laying for one or two months instead of four (Bell and Weaver, 1067).

Molting refers to the replacement of old feathers by new ones. In nature, all birds replace all of their feathers in the course of a year. The process varies within and according to species, although many birds lose and replace most of their feathers in the fall at the onset of the cold season to provide warm plumage for the winter. Egg laying tapers off as the female bird concentrates her energies on growing new feathers and staying warm. Nature discourages the hatching of chicks in winter, when food is scarce.

The U.S. egg industry uses forced molting as an economic tool to regulate egg prices, renew eggshell quality, and reduce the fat that accumulates in the oviducts of unexercised hens (Bell and Weaver, 1074). Forced molting programs are designed to shock hens into losing 25 to 30 percent of their weight "in body fat, feathers, liver tissue, musculature, and skeleton" (Bell 1996, 4). By the tenth to fourteenth day of total food deprivation, a hen who weighed 3.65 pounds before the molt weighs 2.56 to 2.73 pounds ("Kalmbach Feeds"). Such drastic weight reduction does not occur in naturally molting birds. Naturally molting birds do not stop eating, pluck and consume other birds' feathers to reduce hunger, develop systemic and infectious disease conditions, or lose immune function and die in large numbers, as do force-molted birds.

Poultry researchers invent, duplicate, and refine starvation and nutrient-reduction methods in experiments designed for commercial use and to perpetuate the research (Bell 2005). The three main methods of forced molting are (1) elimination or limitation of food and/or water, (2) feeding the birds low-nutrient diets deficient in protein, calcium, or sodium, and (3) administration of drugs and metals including methalibure, enheptin, progesterone, chlormadinone, aluminum, iodine, and zinc (Bell and Weaver, 1067, 1071–1072).

Forced molting involves bizarre blackout and lighting schedules, which increase *Salmonella* colonization of the hens' internal organs as a result of stress on their immune systems ("*Salmonella* Progress"). A "very popular" method, developed at North Carolina State University, includes a week of 24-hour continuous artificial lighting prior to food deprivation for 14 days, followed by a reduction to 10 hours of light per day during the two-week starvation period. The California Molting Program reduces light to 8 or 10 hours during the 7 to 14 days of complete absence of food. "On or about the 7th day without feed the flock will begin to look unthrifty," say North and Bell, who advise not to feed the hens unless mortality reaches unprofitable levels (1069–1071).

Though water removal is no longer recommended, some producers may still "include 1 or 2 days of water removal to help get the flock out of production" (Bell and Weaver, 1067). In addition, the heat is turned up to between 80 and 90 degrees F (26 and 32 degrees C) in the house to reduce the amount of energy the hens would otherwise have to expend to stay warm without their feathers (which, as the molt proceeds, are as thick in the air as a pillow fight, a researcher told me, chuckling) and because high heat decreases their desire to eat (Ernst; Elliot). After the first forced molt, "the house will usually be only 90 to 95 percent full" (Bell and Weaver, 1073).

Food and water deprivation for more than 24 hours was banned in Britain by the 1987 Welfare of Battery Hens Regulations, "except in the case of therapeutic or prophylactic treatment." In 1994, food deprivation to induce a molt was banned in the European Union as both cruel and unsafe ("Community Legislation"). Under Canada's economic system, hens have traditionally been slaughtered after 10 months or a year of egg-production instead of being force-molted and reused for a second cycle (Duncan 1999). This made it easy for the Canadian Veterinary Medical Association to condemn forced molting by food and water deprivation as "sufficiently severe to compromise animal welfare" (CVMA) in contrast to the American Veterinary Medical Association, which didn't concur (UPC 2003c; "Delegates Act") until a full-page ad in *The New York Times*, sponsored by Animal Rights International, United Poultry Concerns, PETA, and the Association of Veterinarians for Animal Rights, exposed the AVMA's violation of its oath to relieve animal suffering (ARI).

Forced molting has been practiced and discussed since the turn of the twentieth century. In *The Molting of Fowls*, published in 1908, researchers in the Department of Poultry Husbandry at Cornell University discussed experiments to "force the molt" by "starving the fowls for a few weeks, which would cause egg-production to cease and the feathers to loosen through lack of nourishment" (Rice, 19). Egg producers began force-molting hens commercially in 1932, in Washington State (Skinner, 709).

In 1967, Donald Bell described nine different forced molting experiments designed to "rejuvenate" the reproductive systems of hens and force them into another laying cycle. He considered it "interesting" that there was a "trend toward more mortality using the severe starvation methods"—no food for 10 days, no water for three (Bell 1967, 24). In 1992, he published an article describing the effects on egg production of starving chickens for 10 or 14 days, followed by restricting their diet for 14 or 18 days. He concluded: "Fasting periods can range from 5 to 18 days, but the use of these extremes should be examined carefully and economic considerations should be part of any such analysis" (Bell and Kuney, 206).

Campaign to Stop Forced Molting by Food and Water Deprivation

United Egg Producers (UEP) has formed an animal welfare scientific advisory committee to review the animal welfare guidelines that UEP developed in 1983—the first such guidelines in the animal agriculture industry. UEP noted that letter-writing campaigns mounted by United Poultry Concerns and [the Association of] Veterinarians for Animal Rights have led to "several thousand" letters sent to UEP's headquarters office in Atlanta, Ga.

—Rod Smith 1999, 9

In 1992, I discovered the literature on forced molting and published "Starving Hens for Profit" in the summer 1992 issue of our quarterly magazine, *Poultry Press*. The following year I found information on the link between forced molting and *Salmonella enteritidis* in hens ("Cruelty and Salmonella"). These discoveries prompted a campaign that lasted for more than a decade in which United Poultry Concerns and the Association of Veterinarians for Animal Rights(AVAR) worked together to expose the egg industry's practice of starving hens for profit (Davis and Buyukmihci; Davis and Cheever).

Determined to get more information, I mailed an "advance survey to assess the range and benefits of induced molting of layer flocks in the United States" to one hundred U.S. egg producers listed in the annual WATT directory, *Who's Who in the Egg and Poultry Industry*, in 1998. In response, I received signed statements from several producers who said they deprived their hens of food between four and fourteen days at a time (UPC 1998).

That same year, in response to my inquiries, we received letters from the U.S. Department of Agriculture's Food Safety and Inspection Service (Stolfa) and Animal and Plant Health Inspection Service (Reed) acknowledging the link between forced molting and *Salmonella enteritidis* infection in hens and the cruelty of the starvation practice. APHIS wrote: "We understand and share your concerns about the humaneness of this practice as well as the food safety issue. . . . The UPC campaign on the forced molting of poultry will help to raise the public consciousness on the issue and create opportunities for action from interested parties" (Reed). That is what happened.

Armed with the science, UPC and AVAR formally petitioned the Food and Drug Administration and the Department of Agriculture in 1998 to prohibit forced molting-induced starvation ("Citizens Petition").

At the same time, we generated "more than 5,000 cards, letters, and signed petitions to the offices of the United Egg Producers in Atlanta calling for the egg industry to discontinue its practice to force hens to molt" ("UEP Plans Research"). Documents obtained by United Poultry Concerns in 1999 through a Freedom of Information Act request confirmed that the government had known for years that starvation-induced forced molting significantly increased *Salmonella* infections by weakening hens' immune systems (Marquis).

In 1999, I urged *The Washington Post* journalist Marc Kaufman to do a story about forced molting based on the Freedom of Information Act evidence I'd obtained. The break came when California assemblyman Ted Lempert introduced Assembly Bill 2141 (which got a hearing but died in the agriculture committee) to ban forced molting in California. Kaufman's article "Cracks in the Egg Industry: Criticism Mounts to End Forced Molting Practice" appeared on the front page of the Sunday edition of *The Washington Post* on April 30, 2000.

Overwhelmed with criticism—poultry scientists Ian Duncan and Joy Mench spoke out—the egg industry decided that instead of hens being starved until their combs turn blue, they could be fed a "molt diet" consisting of wheat middlings, corn, and other low-protein, low-calcium ingredients to obtain the same results (Smith 2003a, Pope).

In 2004, United Egg Producers surveyed 46 companies to determine what kind of a molting program producers were using. In response, 22 companies said they used a food-deprivation program and 24 said they were using a low-nutrient molt diet of some kind. Eighty percent of the latter companies said they would continue using the molt diet rather than food deprivation. In February 2005, United Egg Producers amended its *Animal Husbandry Guidelines for U.S. Egg Laying Flocks* stating that after January 1, 2006, "only non-feed withdrawal molt methods will be permitted" (Dudley-Cash 2007).

Disposal of Spent Hens

When I visited a large egg layer operation and saw old hens that had reached the end of their productive life, I WAS HORRIFIED. Egg layers bred for maximum egg production were nervous wrecks that had beaten off half their feathers by constant flapping against the cage.

—Temple Grandin 2001

Spent hens reach the end of their laying cycle in a physically fragile state.

—"Spent Hen Disposal Across Canada," 2003

Egg companies typically do not see themselves as being responsible for a hen's welfare once ownership of the bird is transferred, and what happens to spent hens after that tends to fall under the radar.

—Bruce Webster 2007

Before going to slaughter, laying hens are deprived of food for an average of four days to "provide a modest net return to help pay for the costs of hen disposition. . . . The greatest benefit of fasting occurs on the third day. In this scenario, fasting a flock provides as much as 3.6 cents extra per hen that can be put against the cost of flock removal" (Webster 1996).

The hens are trucked to slaughter in wire cages or plastic crates without food or water for hundreds of miles, frequently across state lines or into Canada, often with missing legs, feet, and wings that were left behind during catching. Hens who are still laying eggs are pasted in egg slime and pieces of shells. Hens who escape during catching are brutally rounded up from the conveyer belts and manure pits in which they take refuge. A witness described this savage hen hunt at a complex in Mississippi. The hens were pulled out of their hiding places by the neck-breakers who killed or half killed them and loaded them into dump trucks "piled as high as would allow without pouring bodies over the sides" (Stanley-Branscum 1995).

An article in the *Vancouver Courier* conveys some of the horror in which hens travel to slaughter: "Stuffed in cages stacked on top of each other, the birds rain urine and feces down on each other, and the discomfort doesn't stop there. In the United Kingdom a study found that 29 per cent of the 'spent hens' arriving for slaughter had broken bones" (Miller 2001, 3). A U.K. study found that the common method of grabbing hens from the battery cages by one leg each increased bone breaks in their legs and bodies to 13.8 percent over 4.6 percent when hens were grabbed from cages by both legs (Knowles).

At slaughter, spent laying hens are a mass of broken bones, abscesses oozing yellow fluids, bright red bruises, internal hemorrhaging, and malignant tumors (Devine; Druce 1989, 11–12). They've lost 40 percent or more of their feathers (Elliot), and because they are economically "worthless," they sit in the transport cages in all weathers at the slaughterhouse "until all other birds are dealt with—up to 12 hours" (Duncan 2006). The slaughtered birds are shredded into products that hide the true state of their flesh and their lives: chicken soups, pies, and nuggets, commercial mink and pet food, livestock and poultry feed, and school lunches and other institutionalized food service and government purchase programs developed by the egg industry and the Department of Agriculture to dump dead laying hens onto consumers in diced up form (Brown 1990, 1992, 1997; Dickens; "Alternatives"; "Spent hens").

"Collapsed Spent-Hen Market"

Calls to television stations and letters to newspapers indicate that Americans were mostly shocked by coverage of live burial and sometimes live incineration of chickens in Southeast Asia to stop the spread of avian flu H5N1—but live burial of chickens is also common here, to dispose of "spent" hens and surplus male chicks from laying hen "factories."

—Merritt Clifton, *Animal People*, March 2004

In Newfoundland, hens are now killed on-farm and transported for processing as a food source for farmed foxes. "It is a win-win situation for both industries."

—"Spent Hen Disposal Across Canada," 2003

With little flesh on their bones compared to spent "meat-type" breeder chickens, and because there are many millions of them—for example, 54 million spent commercial laying hens (referred to as "light" chickens) were reported slaughtered in the U.S. in 2006 (USDA-NASS 2007, 1), and about 40 million are slaughtered in Britain each year (Kristensen 2001)—spent laying hens have no market value. For this reason, explained *Feedstuffs* in 1997, an increasing number of spent laying hens are being disposed of "outside traditional channels" in on-farm processes that involve "burial or composting or are being rendered" (Smith 1997). The hens are suffocated in large dumpsters and then picked up by the rendering plant to be turned into animal feed ingredients (Gross).

Gas-Killing and Live Burial

An article in 2006 said "there isn't a California facility willing to take them." With a half million hens to be gotten rid of each year in Sonoma County alone, many of these hens are being crammed into sealed boxes filled with carbon dioxide and then composted, dead or alive. Farmers call hens who crawl out of the compost in which they were buried under mounds of sawdust "zombie chickens" (Young).

A common method of destruction, according to Tom Hughes in Canada, is "to pack the birds into a closed truck and connect the exhaust to the body of the truck" (Clifton 2000). On-farm gassing of spent hens by the egg industry has been under investigation in the United States, Canada, and elsewhere for over two decades. In 2005, Alberta Egg Producers announced support for a "cheap, efficient," and "very humane" system of destroying large numbers of hens by dumping them in batches into deep bins designed to hold 650 birds at a time, and pumping carbon dioxide into the bins (Duckworth).

Claims that CO_2 gassing produces a humane death conflict with scientific evidence showing that carbon dioxide induces a breathing distress in mammals and birds known as dyspnea, which "activates brain regions associated with pain and induces an emotional response of panic" (Raj 2004). An article titled "Dyspnea and Pain" begins: "There are few, if any, more unpleasant and frightening experiences than feeling short of breath without any recourse" (Banzett and Moosavi).

CO_2 has been shown to cause extreme suffering in mammals and birds, whereas the inert gas nitrogen is said to be "not expensive" and "much more humane than CO_2" (Duncan 2006; Raj, 2004). However, the egg industry is not interested in the use of inert gases such as argon or nitrogen. One problem with any kind of gas is how to distribute it evenly through the house, into the upper tiers of cages, and keep it there, without oxygen seeping in, until all the birds are dead. Another problem is that the "incoming gas is very cold and has a high speed," according to veterinarian Lotta Berg (UPC 2006). Veterinarian Holly Cheever notes that regardless of the type of enclosure, "if liquid CO_2 is used, the possibility of birds freezing to death before loss of consciousness is high" (Hawthorne).

Woodchippers

In 2003, workers at the Ward Egg Ranch in Southern California threw more than 30,000 spent hens into woodchipping machinery, in which a piece of wood is "fed into a chipper's funnel-shaped opening, and the blades on a rapidly spinning disk or drum cut it into small pieces"

(Fitzsimons). In affidavits obtained by the San Diego County Department of Animal Services and made available to United Poultry Concerns through a Freedom of Information Act request, a plant manager told county investigators that "chipping" chickens is "very common, and tree chipper rental places even advertise in the poultry industry for that use." He said it was easier for the staff to "cram the chickens in a shoot than to chase them around and break their necks" (Carlson 17). A government-industry document included in the file advises egg producers that "whole birds can be composted on site." "Grinding birds first aids composting considerably. We have Southern California farmers who have done this very successfully. A large chipper can be rented and set up to discharge directly into a loader bucket or other container" (Carlson 23).

The woodchipper incident showed the direction of live hen disposal: destroy the birds at the production facility instead of trucking them to slaughter. At a National Carcass Disposal Symposium in 2006 there was "a minor discussion on the 'wood chipper' method and apparently some don't get why that practice [accepted, for example, in Denmark] is not widely accepted" (Guite).

Woodchipper-like machinery, called macerators, are said to be commonly used in Canada to dispose of spent hens (Francois). A Canadian publication in 2003 listed on-farm methods of hen disposal in the provinces to include mobile electrocution units, CO_2 gassing, cervical dislocation (neck-breaking), and macerators. Researchers at the Nova Scotia Agricultural College are experimenting with the "macerator option" to convert spent hens into feed for the fur farm industry. The unit "vacuums" the hens down a tube to a grinder that kills them with blades ("Spent Hen Disposal Across Canada").

Firefighting Foam

The idea is to use a water source and mix it with the same kind of chemicals that firefighters use when fighting wildfires. This mixture goes through a pump apparatus that makes the foam that enters the poultry house. The foam can rise to 4 feet. They claim the birds suffocate and die within 20 minutes.

—Lauren Guite, Food & Water Watch, 2006

FOAM ALONE is what you need. One person can do whole-house depopulation with Avi-Guard.

—Advertisement, *International Expo Guide '07*

In November 2006, the U.S. Department of Agriculture announced its approval of firefighting foam to kill chickens and turkeys in cases of infectious disease outbreaks, such as avian influenza, or when poultry buildings are damaged by disasters, such as hurricanes (AP 2006). USDA said water-based foam can be an alternative to traditionally used carbon dioxide to suffocate floor-reared birds—that is, birds raised for meat as well as cage-free and free-range birds who are on ground-level surfaces as opposed to being stacked in cages.

Earlier in the year, the USDA's Animal and Plant Health Inspection Service held a meeting on methods of mass depopulation of poultry and asked that recommendations be submitted to the agency by nongovernmental organizations attending the meeting (UPC 2006). Describing small trials using firefighting foam to exterminate cage-free hens, Bruce Webster of the University of Georgia presented slides showing "a lot of escape behavior for four to six minutes. You saw the birds' heads sticking out of the foam." Eventually they were "worn out" with their "volitional struggle" (UPC 2006).

Ian Duncan of the University of Guelph in Ontario states, "Foam is a horribly inhumane way to kill birds. You can't tell if they are suffering or vocalizing because they are covered up" (Duncan 2006). In a letter to the USDA, veterinarian Holly Cheever wrote on behalf of the Association of Veterinarians for Animal Rights, "By virtue of their being hidden from view and possibly unable to vocalize as they are covered with the foam, determining their degree of suffering becomes problematic. Also, although the birds do not seem to struggle as the wall of foam approaches them, their immobility should not be interpreted as a lack of stress or concern on the part of the birds. Finally, a board certified veterinary toxicologist states it is likely the chemical ingredients of the foam will cause irritation of the birds' eyes, mucous membranes, and skin" (Hawthorne, 41).

In a report to the U.S. Department of Agriculture on behalf of The Humane Society of the United States, Mohan Raj of the University of Bristol in England wrote: "A primary welfare concern with this method is that the birds appear to be killed either by suffocation or drowning, which is prohibited by law in many developed countries. A physiological definition of suffocation is the physical separation of the upper respiratory tract from the atmospheric air, which would happen if the birds were buried alive. In contrast, it is claimed that the foam induces cerebral anoxia. This may be true, but the method of inducing this physiological state is unacceptable on ethical and humanitarian grounds" (Raj 2006a).

The Fight for Better Conditions

Each individual life we save means the world to us and to them. Pure bliss is watching a withered, featherless, debilitated, and naked little hen look up at the sky for the first time in her life, stretch her frail limbs, and then do what all hens adore: take a dust bath!

—Patty Mark, "To Free a Hen," *Animals' Agenda*, 2001

It will be a week today since we rescued 78 "spent laying" hens. Yesterday one of the hens lay down by the water provided for her and relaxed her wings. I prepared a small animal taxi with soft bedding, attached a container with food and water, and placed her in the warm utility building which is home to our small rescued animals. This morning I found her dead. I grieved, and wanted to share my grief; but I didn't. I did not want to hear, "She's better off now. She won't suffer anymore." While we say these things to console ourselves, and it is true she is not suffering, she died needlessly. She lived a miserable life and she died a miserable death. She was cold and covered with excrement when we found her in the waste pit beneath the battery cages. Her body temperature had dropped and she had not recovered.

The six days she had in the fresh air and sun were not days spent flexing her feet and stretching her wings. These days were likely hazy. Yes, she was warm, food was available to her and she ate and drank, but she died last night after a very, very long short life of suffering. And It's Not Okay. May she burn in the forefront of our minds. May we resolve to expose her life and death.

—Doll Stanley-Branscum, Grenada, Mississippi, 1995

United States

No federal laws protect chickens in the United States, and state anti-cruelty laws typically exempt farming practices that would be illegal if the animals were companion dogs and cats (Wolfson). Birds are excluded from the Animal Welfare Act and from the Humane Methods of Slaughter Act. An example of how laying hens are treated in the U.S. was caught on videotape in 1993 and 1994 during undercover investigations at Boulder Valley Poultry Farms in Colorado (RMAD). *Raw Footage, Raw Pain* shows scenes of hens packed eight to a cage amid the incessant din of bird cries and machinery. Hens are left to die in a closed wing of one of the sheds; piles of dead chickens and chickens with open sores appear among decaying broken eggs and mounds of uncovered manure. A stray hen walks over a pile of dead chickens, and rats whistle through cages in which claws and other body parts of former inmates lie rotting. Whole cages are shown full of dead hens in various stages of decomposition. A veterinarian from the Colorado Department of Agriculture tells a television reporter blandly that this is normal business practice. It still is, but some important changes have meanwhile occurred.

An important turning point in the 1990s was United Poultry Concerns' Forum on Direct Action for Animals on June 26–27, 1999. Australian activist Patty Mark introduced U.S. activists to the concept of open rescues, in which undercover investigators admit to rescuing animals and documenting the conditions of their abuse (Davis 2004). Inspired by the Australian model, three investigations of battery-caged hen facilities were conducted in the United States in 2001. Compassionate Action for Animals openly rescued eleven hens from a Michael Foods egg complex in Minnesota; Compassion Over Killing openly rescued eight hens from ISE-America in Maryland; and Mercy for Animals openly rescued 34 hens from DayLay and Buckeye egg farms in Ohio. These investigations, including photos, videos, and press conferences, drew coverage from *The Washington Post*, Ohio Public Radio and many other media outlets.

Another milestone was reached when I persuaded Paul Shapiro, co-founder of Compassion Over Killing, to join me in attending the WATT Poultry Summit on "Focusing on Bird Welfare in the Commercial Layer Industry" on October 16, 2001, in Henderson, Nevada. Charles Olentine, editor of *WATT PoultryUSA*, wrote in the December 2001 issue of the magazine: "You can imagine our surprise and apprehension when two of the first three registrations came from United Poultry Concerns and Compassion Over Killing. At the time of the announcement, the publicity was targeted at the egg industry itself. As it turned out, the summit brought together over 100 attendees ranging from the hard-line egg producers who do not want to change anything in terms of production practices to groups promoting the vegan way of life" (14).

The summit was "a wake-up call," Olentine wrote, citing the civility, dedication, and knowledge displayed by the activists. He urged producers not to "underestimate the opponent" but to convince consumers that "we care for our birds and that they are not treated as inanimate objects" ("like cabbages," as poultry scientist Ian Duncan in 2006 said they are treated).

So impressed was Olentine with the activist presence at the laying hen welfare summit that he devoted 13 pages of *Egg Industry* magazine in October 2002 to presenting our views and profiling our organizations, including United Poultry Concerns, Compassion Over

Killing, The Humane Society of the United States, and People for the Ethical Treatment of Animals. He interviewed me at our sanctuary in Machipongo, Virginia, where he saw our hens sunning themselves and enjoying their lives.

United Egg Producers heeded the challenge "to show that it views its birds as more than machines" (Olentine 2002, 26). In addition to publishing a revised (from 1983) *Animal Care Guidelines* in 2002, UEP unveiled an Animal Care Certified program in which participating farms would be audited for compliance with the guidelines' "science-based" standards for cage space per hen, air quality, beak trimming, molting, handling, and transportation. Not surprisingly, the vast majority of these farms passed the audit, allowing them to display the Animal Care Certified logo on their egg cartons (McGuire).

False Advertising Claim Upheld

The "Animal Care Certified" stamp on the grocery store egg cartons declared that the chickens were raised in humane conditions, but the tapes tell a different tale.

—Nelson Hernandez, *The Washington Post,* Oct. 4, 2005

Undercover investigations by Compassion Over Killing in Maryland, Mercy for Animals in Ohio, and Compassionate Consumers in New York showed the Animal Care Certified program and logo to be bogus. Nathan Runkle, executive director of Mercy for Animals, explained the fraudulence of Animal Care Certified companies such as Ohio Fresh Eggs (formally Buckeye Eggs). "It's really just a public relations gimmick by the egg industry to play off the good intentions of consumers," he told reporters (Drew).

In 2003, Compassion Over Killing filed a false advertising complaint with the Better Business Bureau, whose National Advertising Review Board, seeing the cruelty being marketed as "humane," recommended that the Animal Care Certified seal be dropped or changed. Facing an investigation by the Federal Trade Commission, the egg industry agreed, in 2005, to replace its Animal Care Certified logo on store cartons with one stating "United Egg Producers Certified" and "Produced in Compliance with United Egg Producers' Animal Husbandry Guidelines." In 2005, the Federal Trade Commission announced that egg cartons with the old logo had to be out of stores by April 1, 2006 (Hernandez).

Wegmans: In November of 2005, our egg farm participated in its annual audit of the United Egg Producers (UEP) Certified program. . . . The USDA conducted this audit, and we received a perfect score—200 out of 200.

—Wegmans Consumer Affairs, March 22, 2006

An example of the clash between the reality of battery-caged egg production and the egg industry's Animal Care Certified logo was provided by Compassionate Consumers. In 2004, they videotaped conditions inside one of the battery-caged facilities owned by Wegmans Food Markets, a 68-store, family-owned supermarket chain based in Rochester, New York, with stores in New York, Pennsylvania, New Jersey, and Virginia.

Wegmans at first tried to argue that the scenes in the video, *Wegmans Cruelty*, didn't come from their facility but was forced to admit as much in order to press charges against Compassionate Consumers investigator Adam Durand, who was subsequently acquitted by a jury of felony burglary and petit larceny in 2006 and sentenced to six months in jail for misdemeanor trespass (York).

Undercover investigations conducted at two Lancaster County, Pennsylvania, operations—Esbenshade Farms and Kreider Farms—revealed the same appalling conditions. The Esbenshade investigation, led by Compassion Over Killing, resulted in 70 counts of criminal animal cruelty against the farm owner and the manager, who were acquitted by a Pennsylvania court in 2007 (COK 2007), and the Kreider Farms investigation, led by Philadelphia-based Hugs for Puppies, refuted the company's website claim that its three and a half million hens were "happy and well treated." In response, Kreider changed the word "happy" to "contented." (Burns).

"Cage-Free"

In 2005, The Humane Society of the United States(HSUS) launched a campaign to get corporations and universities to agree to use only cage-free eggs in their dining facilities. Cage-free means that, while confined in buildings without access to the outdoors and almost always debeaked, the hens are not in wire cages. In 2006, more than 80 schools, including

Dartmouth College, University of Connecticut, University of Iowa, and American University, "enacted policies to eliminate or greatly decrease the use of eggs from caged hens," according to an HSUS press release in April of that year. AOL joined forces to be followed by Google and Ohio State University. Other universities and companies have made similar announcements, including Hellmann's Mayonnaise in 2008.

In the legislative arena, following a campaign by HSUS and sponsored by other pro-animal groups, including United Poultry Concerns, 63 percent of California voters supported the state's Prevention of Farm Animal Cruelty Act, or Proposition 2, on the November 2008 ballot. The law, which goes into effect in 2015, requires California's 20 million egg-laying hens, as well as pregnant sows and calves raised for veal, to be able to stand up, lie down, turn around, and fully extend their limbs without bumping into other animals, bars or walls in the confinement area. This law could eliminate California's battery-cage system of egg-production, in whole or in part, and is expected to bring similar laws to other states (*Feedstuffs* News Flash, Nov. 5, 2008).

In addition, an effort is being made to establish minimum welfare standards for producers, whose compliance would allows them to advertise their eggs as "humanely produced." There's the National Organic Program administered by the U.S. Department of Agriculture, the Certified Humane Program administered by Humane Farm Animal Care, Free Farmed Certified administered by the American Humane Association, Animal Compassionate administered by Whole Foods Market, and Animal Welfare Approved administered by the Animal Welfare Institute. AWI provides an Animal Welfare Approved seal to the independent family farms it works with (www.AWIonline.org).

Canada

Government and industry are constantly reassuring consumers that things are better for farm animals in Canada. We have long suspected that's not the case and now we have the proof—this footage shows filthy, disgusting, hideously abusive conditions.

—Debra Probert, Executive Director of the Vancouver Humane Society, *Canadian Press*, Oct 13, 2005 (Baumel)

The footage referred to above was obtained by an anonymous University of Guelph biology student in 2005 at LEL Farms, a battery-caged hen facility (barn) near Guelph, Ontario, owned by Lloyd Weber, a veterinarian and a member of the Dean's Veterinary Advisory Council of the University of Guelph. This was the first time such footage had been made public in Canada (Probert), and it showed that rather than being an exception, it was simply "battery-hen reality" (Baumel).

The battery cage is legal in Canada, where 98 percent of the country's 26 million egg-laying hens are confined in barns averaging 17,000 birds per barn (Baumel). In 1980, the Canadian Federation of Humane Societies began coordinating voluntary codes of practice for livestock, including poultry from the hatchery to the slaughterhouse, with financial support from the Canadian Food Inspection Agency.

The *Recommended Code of Practice* formulated by the Canadian Agri-Food Research Council (a nongovernmental organization funded by government and industry) was published in 1983 and extended in 1989 and 2003. The 2003 *Recommended Code of Practice for the Care and Handling of Pullets, Layers and Spent Fowl*—nonmandatory in every province but New Brunswick, Prince Edward Island, and Manitoba—allows producers to give each White Leghorn hen 67 square inches per four-pound bird—an addition of three inches to the *Code's* former 64 square inches. The slightly heavier brown hens get 75 square inches per hen (Duckworth 2003). The Canadian Egg Marketing Agency (CEMA) has an Animal Care Certification Program similar to that of United Egg Producers in the United States.

Canada's Ban Battery Cages Campaign

The Canadian Coalition for Farm Animals and the Vancouver Humane Society have posted "The Truth about Canada's Egg Industry," based on the undercover footage obtained at LEL Farms in 2005, on their websites. They've asked the Canadian Food Inspection Agency to adopt the European Union's policy of making retailers label all battery eggs as "eggs from caged hens" and to stop using terms like "farm fresh" and "natural." They're also campaigning to get food sellers to make these labeling changes voluntarily and to sell at least 50 percent of their eggs from uncaged hens. These campaigns and how to get involved can be viewed online at chickenout.ca, humanefood.ca, and aquarianonline.com/guide.htm.

European Union

"A New Era of Humanity for Hens"?

In the European Union, approximately 400 million hens—88 to 90 percent of commercial laying hens in the EU—are kept in battery cages typically holding four hens per cage (CIWF Trust 2004). In the mid-1990s, the future of battery cages was under discussion in Europe, following the production of draft proposals for a directive by the European Economic Community Commission. A 1992 report from the Commission's Scientific Veterinary Committee concluded that the barren battery cage did not provide "an adequate environment or meet the behavioural needs of laying hens" (EECC).

On June 17, 1999, the European Union announced Laying Hens Directive 1999/74/EC. The Directive effectively banned the barren battery-hen cage in Europe by 2012 by adopting the Swiss formulation of minimum conditions, which cannot be met by conventional cages (Studer, 46). Compassion in World Farming, which has fought for decades to ban battery cages, hailed the decision as "a new era of humanity for hens."

Until 2012, existing cage systems are required to be slightly improved by reducing the number of hens per cage so that each hen has 86 square inches of living space, up from the current 70 square inches. In 2008, the European Commission reaffirmed its directive banning conventional cage systems in the EU in 2012 (Smith 2008).

"Enriched" Cages

A problem with the EU's 2012 ban is that it allows the use of so-called enriched cages. An "enriched" cage has a tiny perch and nest box and maybe a little bit of litter (sand or wood shavings) for pecking, scratching, and dustbathing. The crowded hens have "extra" space the size of a postcard. Thousands of teensy "sandboxes" will only increase the airborne debris in the caged environment already mired in filth. In addition, Farm Animal Welfare Network (FAWN) has pointed out that while a nest box is vital to the welfare of hens, the "enriched" cage will make meaningful inspections, which are already next to impossible, even harder—inspections are required in England by The Welfare of Farmed Animals Regulations 2000—while retaining the inherent cruelty of the cage. FAWN's founder and director, Clare Druce, asks, "Will the nesting box be carefully inspected, daily? Will checks be made to see if a hen in there is in fact laying [an egg], resting, escaping from bullies, or merely dying from cage layer fatigue?" (Druce 2006).

Compassion in World Farming and FAWN remain concerned that the egg industry's effort to delay the 2012 ban by up to 10 years or longer could succeed, along with the fact that "there is little legislation to address minimum standards of hen welfare" once the ban is enacted. In *Practical Alternatives to Battery Cages for Laying Hens: Case Studies from Across the European Union, 2004,* CIWF Trust lists standards the group says need to be adopted and implemented, including the use of hardy breeds of hens suited to free-range (outdoor) conditions; the banning of beak trimming, which is outlawed in Sweden; outdoor runs that include overhead cover such as trees; materials suitable for nesting, foraging and comfort; installation of perches; and lower stocking densities to promote natural behavior and reduce feather pecking. *Practical Alternatives* looks at nine free-range and organic laying-hen operations in Sweden, Spain, France, Belgium, and the U.K. It's available from Compassion in World Farming along with their 2006 report *The Way Forward for Europe's Egg Industry: Keeping the Ban on Battery Cages in 2012.*

SWITZERLAND

Switzerland didn't generally prohibit batteries. It simply defined higher standards and hasn't authorized any more cage systems.

—Heinzpeter Studer, *How Switzerland Got Rid of Battery Hen Cages*

Barren battery cages for laying hens were installed in Switzerland in 1935 and abolished in 1991, effective January 1, 1992. The story of how this happened is told by Heinzpeter Studer in his 60-page book, with color photographs, *How Switzerland Got Rid of Battery Hen Cages.* Published in German in 2001, the book has been translated into English by United Poultry Concerns and can be found on our website at www.upc-online.org/battery_hens/SwissHens.pdf.

Starting with a Swiss Animal Welfare Act in 1978 that set basic welfare standards for hens, with the approval of 81 percent of the Swiss popular vote despite fierce opposition from egg producers, the campaign to get rid of battery cages led to a 1981 Animal Welfare Ordinance stipulating that "hens must be provided with at least 800 square centimeters [124 square inches] accessible floor area per bird, with protected and shaded nests as well as with perches." Clearly, the battery-cage system could not meet these standards.

Swiss animal protectionists then launched a massive public relations campaign that drove battery cages out of Switzerland, although imports from other countries and the use of battery eggs in processed foods continue. Overcoming its initial opposition, the Swiss Egg Producers Association chose to work with the two biggest Swiss food sellers, Migros and Coop, along with animal and environmental activists and the Swiss government, to create a positive image of the Swiss consumer—proud to pay a little more money for a healthier, more ethically obtained egg.

While the 1991 Swiss law prohibits barren battery cages, it does not prohibit so-called enriched or furnished cages. However, the expense of maintaining and cleaning such cages is so high that they haven't become established in Switzerland. A concern, however, is that as a member of the European Union, Switzerland would lower its standards to those of the European Union. As the EU doesn't stipulate the official inspection and authorization procedures that have kept barren battery cages out of Switzerland, an EU "detour via more expensive 'enriched cages' could also lead back to the old batteries," Studer warns.

Fear of the EU's influence is justified given that the "egg producing establishments elsewhere in the EU are much bigger than those in Switzerland," he says. For example, "If somebody has 200,000 hens in the battery who are cared for by one and a half workers, then it is obvious that this doesn't work in an aviary [a building with additional levels of platforms and perches interconnected by ladders, and generally considered the best husbandry system for combining the basic needs of hens with economics], because an aviary doesn't work with so little time for each animal" (Studer, 48–50).

It appears that the fate of hens in Switzerland will depend very much on what happens in the European Union. According to a 2005 government report, EU egg producers claim that the ban will greatly increase the cost of doing business and promote "bacteriological, health, and welfare problems" among free-range and non-caged hens. At the same time, however, "animal welfare aspects of animal farming are becoming more important, not only in Europe but also in the United States and other parts of the world" (Francis).

Sweden and Germany

Sweden banned battery cages in 1999. In 2005, the European Commission began a legal process to try to invalidate Sweden's law as hindering trade among EU member states ("Local"). In Germany, conventional barren cages were to be banned from January 2007, and enriched cages from 2012. However, in 2006, Germany's government accepted a parliamentary proposal to postpone the decision of its former agriculture minister to ban all cages for laying hens in Germany by 2007. German egg producers now have until the end of 2008 to change from standard cages to small aviary systems, each holding 35 to 50 hens ("Germany"), in which hens will have floor space amounting to about one square foot per bird, the standard allotment in virtually all non-cage systems, although aviaries also have perches and platforms on which to crowd more hens, which many non-cage systems do not.

Australia and New Zealand

The battle over battery cages in Australia and New Zealand is summarized in Heinzpeter Studer's book *How Switzerland Got Rid of Battery Hen Cages* online at www.upc-online.org/battery_hens/SwissHens.pdf. The publication of *Animal Liberation* by Australian philosopher Peter Singer, in 1975, resulted in animal liberation organizations being formed in all of the six states and two territories of Australia. Banning battery cages was a major goal. In the 1980s, the Senate Select Committee on Animal Welfare reviewed the idea, concluding in 1990 that a ban was feasible only if viable alternatives could be developed. A plan to give caged hens 93 square inches of space per hen (600 square centimeters) under a revised voluntary national Code of Practice for poultry-keeping was gutted by the egg industry, and a national regulatory minimum of space per hen was set in 1993 to 1994 at 69 3/4 square inches (450 square centimeters).

In 1997, the government of the Australian Capital Territory (ACT) passed legislation to abolish the battery cage within six years, but politics undermined the legislation.

In August 1999, the Agriculture and Resource Management Council of Australia and New Zealand (ARMCANZ) considered a national ban on battery cages, but disputes among the states and territories and pressure by the egg industry—which in 2000 had agreed on a cage ban after 15 years—stalled the effort to the point of forcing the umbrella welfare organization Animals Australia to stop its media campaign.

At the ARMCANZ session in August 2000, Tasmania, Queensland, and ACT opted to abolish battery cages. However, New South Wales and Victoria opposed it, other state agriculture ministers abstained from voting, and ARMCANZ ended by guaranteeing that battery cages may continue to be used for up to 20 years after their installation, with new cages required merely to give each hen a minimum area of 85 1/4 square inches (550 square centimeters) inside the cage. In 2001, ARMCANZ agreed to label eggs from battery-caged hens "cage eggs," overriding animal welfare demand that these eggs be more explicitly declared "battery cage eggs" (Studer, 53–54).

In the first edition of this book, I described the campaign waged in the early 1990s by Australian activists Pam Clarke and Patty Mark of the organization Animal Liberation, which resulted in a successful prosecution against Golden Egg Farm in Tasmania. On February 24, 1993, Magistrate Philip Wright found Golden Egg Farm guilty on seven counts under the Tasmanian Prevention of Cruelty to Animals Act 1925. He delivered a historic 18-page judgment against the farm and the battery system, ruling that the hens purchased from the farm had lived in chronic pain from being forced to rub against cage wires. He said that confinement causing the state of the hens submitted in evidence "could not be called other than cruel in my opinion: if a bird is unable to move without affecting, physically, others in the cage, nor to lay [eggs] or rest without affecting itself deleteriously, the cruelty is constant and continual and without relief."

Although debeaking was not the issue in this case, Wright condemned it. Abnormal growths, pus, and ulcers in the beaks of several birds were medically entered in evidence. He condemned the whole system: "The only evidence in this case referring to justification or necessity for the cruelty inflicted upon these birds was in the broadest terms as to economy and profitability of egg production, but such references by no means deflect me from what otherwise would be and is my strong view that all these birds have been treated with unjustified and unnecessary cruelty, constituted by great indifference to their suffering and pain" (*Clarke v Golden Egg Farm*).

Shortly after the ruling, Pam Clarke was charged with trespassing at Hobart Parliament House where she was protesting government efforts to amend the Tasmanian Prevention of Cruelty to Animals Act 1925 to exempt poultry from protection, which would allow the government to override the court judgment ("Pam Clarke"). She was sent to prison for three weeks, among her many incarcerations on behalf of battery-caged hens.

At present, animal activists throughout Australia and New Zealand are fighting for a ban on battery cages. The undercover raids, private prosecutions, and media exposure of the atrocities to which Australian hens are subjected continue. Patty Mark's group, Animal Liberation Victoria, has been a particularly effective investigative organization and has served as a model for American activists working to eliminate battery cages.

The battle to liberate hens from battery cages has begun, and it includes all of us. Wherever we are, we are morally obligated to end the oppression. Battery cages should be abolished in the United States and throughout the world. Until they have been discontinued, our species stands condemned of a criminal relationship with the living world. People should boycott battery eggs and discover the variety of egg-free alternatives.

Chapter 4

The Life of the Broiler Chicken

The misery of egg-laying birds has been well documented, but what about the life of chickens bred for eating?

—Andrew Purvis, "Pecking Order," *The Guardian*

The most appalling thing we witnessed was a broiler facility that produces chickens for eating. We went in and it was totally dark, just three to four dim lightbulbs. They only vented the facility periodically and the dust and ammonia smells were overwhelming.

—Robert Martin, Executive Director of the Pew Commission on Industrial Farm Animal Production, *E Magazine*

Industrial chicken production "must constitute, in both magnitude and severity, the single most severe, systematic example of man's inhumanity to another sentient animal."

—John Webster, *A Cool Eye Towards Eden*, 156

In the past, eggs were the primary source of revenue for the chicken industry. In 1960, eggs supplied 61 percent of the gross chicken income in the United States followed by broiler chickens at 34 percent. This changed in 1975 when, for the first time, broiler chickens supplied 50 percent of the gross chicken income followed by eggs at 48 percent (Hartman). Since then, broiler chicken sales have dominated the poultry industry in the United States. Of a total producer value of $15 billion for the 1992 marketing year, broiler chickens, eggs, turkeys, and other chickens contributed 61, 23, 16, and less than 1 percent respectively. By the end of the 1990s, the U.S. broiler chicken business was a $22 billion industry compared to a $3 billion table egg industry (Bell and Weaver, 29, 945).

Changing lifestyles and attitudes about health are reflected in these figures. Between 1960 and 1990, egg consumption dropped from 320.7 to 234.8 per person per year reflecting health concerns about cholesterol, growing fear of *Salmonella* food poisoning, and the decline of the big breakfast in the American diet (Pressler; Gyles). By contrast, chicken consumption rose from 42 pounds per person in 1970 to 87 pounds per person in 2007, representing a rise from 23 percent to 43 percent of the meat Americans are now eating, mainly in the form of chicken breasts, fast food, and other convenience items. McDonald's alone purchases 650 million pounds of chicken per year for its worldwide restaurants (Watts).

Development of the Modern Broiler Chicken

The development of the broiler chicken industry from a family enterprise to a commercial agribusiness was featured in a special issue of *Broiler Industry* magazine in July 1976. As the nation celebrated its 200th anniversary that month, the chicken industry celebrated its 50th, according to the editors. Ray Goldberg, who with John Davis coined the term "agribusiness" in the mid-1950s at the Harvard Business School, observed: "One would have been hard-pressed 50 years ago to find even a dozen flocks of chickens in lots of as large as 10,000 per farm that were being raised especially for supplying poultry meat. Most of the nation's flocks averaged 75 birds per farm per year and were kept for eggs, with meat a by-product when the hen was laid out."

"Broiler" chickens—birds raised specifically for meat as opposed to being derived from egg production and eaten—were reportedly raised as early as 1800 in the South Jersey town of Hammonton, in incubators capable of hatching up to 100,000 chicks every 10 weeks. A flock of 500 broiler chickens was reported in 1917 in Gainesville, Georgia, and 7,000 broiler chickens were reportedly raised in Smyrna, Georgia, in 1901 (*Broiler Industry* 1976, 14).

The U.S. Department of Agriculture traces the beginning of continuous year-round production of broiler chickens to Cecile Long (Mrs. Wilmer) Steele in Ocean View, Delaware. In 1923, she raised a winter flock of 500 birds, of which 387 survived for slaughter. The industry sets 1926 as the start of its era. In that year, Steele and her husband built a year-round farm capable of producing 10,000 chickens, and the first railroad car full of live broiler chickens (as opposed to "run-of-the-farm fowl") was shipped from New Hampshire to New York City, at a time when live chickens were being transported from New England to New York by truck and from the Midwest to the eastern markets by train, to be slaughtered in the back rooms of grocery stores.

The broiler chickens of those days were not the pure white, oversized birds subsequently developed at universities and land-grant colleges in conjunction with chemical companies such as Upjohn and Merck. They consisted mainly of the black and white Barred Rock chickens developed by poultry farmers in New England in the early 1800s, and the chestnut-colored New Hampshire chickens developed by poultry farmers from the Rhode Island Red chicken in the early twentieth century. Compared to today's four to eight pound bird slaughtered between six and eight weeks old, these birds weighed less than two pounds when slaughtered at fourteen weeks old.

National Chicken of Tomorrow Program

A major event that led to the development of the modern broiler chicken was the National Chicken of Tomorrow program. This competition, sponsored by a New York City advertising agency, included two three-year contests paid for by the Great Atlantic and Pacific Tea Company (A&P food stores) in 1946 to 1948 and 1949 to 1951. The goal was to develop a "super chicken" by evaluating strains based on meat quality and cost per pound of a 13-week-old bird (Skinner, 399–410).

Launched in Monmouth, Maine, in 1946, with state contests the first year, regional contests the second year, and national contests the third year, the program produced a surprise winner, Charles Vantress, of Marysville, California. Until then, New England had been the main source of breeding stock in the country. Vantress's red-feathered Cornish-New Hampshire cross introduced Cornish blood into broiler chicken breeding, which gave "the broad-breasted appearance that would soon be demanded with the emphasis on marketing that followed the war" (*Broiler Industry* 1976, 56).

Arbor Acres of Connecticut was another big winner. Originally a family-operated fruit and vegetable farm, later a Rockefeller subsidiary, now one of the largest broiler breeding companies in the world, Arbor Acres developed the Arbor Acres White Rock from Plymouth White Rock genetic lines. During the 1950s, white chickens replaced birds with colored feathers. Dark pin feathers on carcasses were considered unsightly and were hard to remove. The modern commercial broiler chicken is a highly specialized hybrid in which Cornish male lines and White Plymouth Rock female lines impart primary characteristics (Skinner, 400–401).

The broiler chicken industry's measure of success is aptly characterized in a brochure aimed at college students published by the Merck pharmaceutical company and the National Broiler Council. Renamed the National Chicken Council in 1999, the U.S. trade group formed in 1955 to stimulate chicken consumption: "Dramatic changes have taken place within the industry. Instead of 'scratching for their food,' today's pampered chickens are the products of advanced science and technology." Students don't have to worry about the facts: "When you choose a career in the poultry industry, you may not see a chicken or an egg or a turkey—except at mealtime" (NBC).

Troubled Birds

The technology built into buildings and equipment [is] embodied genetically into the chicken itself.

—Bell and Weaver, *Commercial Chicken Meat and Egg Production*, 805

These birds are all extremely unfit. There is also the interaction between their unfitness and their poor environment.

—Ian Duncan 2001

My own acquaintance with broiler chickens began in the mid-1980s, when my husband and I rented a house on a piece of land that included a backyard chicken shed in Maryland. One day in June about a hundred young chickens appeared in the shed. A few weeks later the chickens were huge. I knew little about broiler chickens at the time, but I was impressed by how crippled these birds were. I saw what Jim Mason and Peter Singer meant by what they wrote in *Animal Factories*: "Fleshly bodies of broiler chickens . . . grow heavy so quickly that development of their bones and joints can't keep up. . . . Many of these animals crouch or hobble about in pain on flawed feet and legs" (45).

The broiler chicken industry tells the public that thanks to pharmaceutical research, better management, diet and related improvements, poultry diseases have been practically eliminated. However, industry publications and my own experience tell a totally different story. A major part of this story concerns what has been done to chickens genetically to create an obscenely heavy, fast-growing bird, falsely promoted to consumers as "healthy." Another part concerns the environment in which these birds are raised and how and what they are fed. Still another part concerns the miserable existence of the male and female chickens who are used to produce the eggs that become the birds people know as "chicken."

A description in the *Atlanta Journal-Constitution* in 1991 is as true today as it was then: "Every week throughout the South, millions of chickens leaking yellow pus, stained by green feces, contaminated by harmful bacteria, or marred by lung and heart infections, cancerous tumors, or skin conditions are shipped for sale to consumers, instead of being condemned and destroyed." One inspector said: "I've had bad air sac birds that had yellow pus visibly coming out of their insides, and I was told to save the breast meat off them and even save the second joint of the wing. You might get those breasts today at a store in a package of breast fillets. And you might get the other part in a pack of buffalo wings" (Bronstein).

Forced Rapid Growth

Growth-related mortality, thought to be a by-product of the modern broiler's ability to gain weight so rapidly, has become more and more of a problem in recent years. This mortality is seen in the form of leg disorders, ascites, and sudden death syndrome. In cooler climates where growth rates are especially fast, and within companies growing large birds for roaster or further processing markets, it is not uncommon for mortality attributable to these causes to exceed 10 percent.

—Bell and Weaver, 876

These half-day-old chicks weigh less than an ounce apiece. In the next month and a half they will multiply that about 65 times.

—Tom Horton, "42-Day Wonders," 2006

In 1935, the average "broiler" chicken weighed 2.80 pounds at four months old. In 1994, the bird weighed 4.65 pounds at six and a half weeks old, reflecting a 3.5 percent faster growth rate than that of 35 years earlier (Leach). Today, a six- to seven-week-old chicken weighs between 4.5 and 8 pounds (Bjerklie). Already, in the early 1980s, when broiler chickens weighed 4 pounds at eight weeks old—43.7 times their original hatching weight—the U.S. Department of Agriculture bragged that if human beings grew at the same rate, "an 8-week-old baby would weigh 349 pounds" (USDA 1982).

One wonders how people would react to being told that health experts link such horrendous growth rates and body weight to human health and well-being, or that tripling the appetite and number of fat cells in people, and making them consume rich food without exercise, was "healthier." Contrary to the myth of lean cuisine, chickens contain more fat than ever before—as much as 15 percent more than in the 1960s. In 1988, the National Research Council explained:

> Genetic selection for body weight caused chickens with above-average appetites to be chosen as breeders. As a result, broilers were produced that ate more feed at a given age and became unable to synthesize protein and lean meat fast enough to keep pace with the increased intake of food energy. The excess food energy was deposited as lipids, and broilers became fatter (Gyles, 299, 307).

Researchers at the Institute of Brain Chemistry and Human Nutrition at London Metropolitan University in 2005 found that modern broiler chickens contain nearly three times the amount of fat that they did 35 years ago. They attributed this gain to the fact that whereas chickens used to "roam free and eat herbs and seeds," they are "now fed with high-energy foods and even most organic chickens don't have to walk any distance to eat" (Ungoed-Thomas). Lack of exercise is part of the story, but the other part of the story is genetics.

For the poultry industry, the challenge is how to continue increasing the birds' growth rate and body weight while managing the poor health and lack of fitness that accompany the increase. Despite warnings by poultry scientists that "broilers now grow so rapidly that the heart and lungs are not developed well enough to support the remainder of the body, resulting in congestive heart failure and tremendous death losses" (Martin), the National Chicken Council insists the industry will continue to raise birds to heavier weights and "process increasingly larger birds and breast meat production into appropriate portion sizes for consumers" (Smith 2003b). Researchers at the University of Guelph in Ontario predict that by 2012, chickens who are now heavy enough to be slaughtered at 35 days old will be ready for slaughter in 26 days (Brown 2007).

A heavy person suffering from painful arthritis and heart disease, without the relief of medication, can easily imagine how these chickens must feel. Reluctant to move, the birds sit heavily in positions that cause their feet, hocks (the top joint of the leg), and breasts to exert tremendous pressure on the ammoniated floor covered with damp bedding and droppings. In time, the ammonia from the decomposing uric acid in the droppings burns into these sensitive pressure areas causing ulcers to form on the birds' feet and blisters to form on their legs and breasts, similar to bed sores.

Bacterial Rot

These wounds invite bacteria. Bones, tendon sheaths, and leg joints become infected with bacterial agents such as *Staphylococcus aureus, Clostridium perfringens,* and *Clostridium septicum.* Ulcerative and necrotic diseases in chickens and turkeys raised for meat are a worldwide phenomenon (Bilgili 2008). An example is femoral head necrosis, a musculoskeletal disease in which the top of the leg bone disintegrates as a result of bacterial infection, excess body weight, and oxygen deficiency (Purvis; Aiello).

The disease known as gangrenous dermatitis affects the legs, wings, breast, vent area, abdomen, and intestines in birds between two and eight weeks old as a result of toxins produced by *Clostridium perfringens* in conjunction with filth and exposure to immunosuppressive viruses. Researchers describe the disease in terms of "moist raw or dark areas" where the underlying muscles are exposed, "blood-tinged fluid of jelly-like" consistency beneath the skin, and livers and spleens "swollen and dark with spots or blotches."

Another prevalent disease caused by *Clostridium perfringens* is necrotic enteritis. Chickens suffering from necrotic enteritis cannot digest their food. They suffer intensely and die a painful death. Their intestines swell up with gas and a foul-smelling brown fluid and show "ulcers or light yellow spots" (Clark 2004; 2007). In 2007, an Oklahoma chicken farmer told me that his chickens were "all rotting in their insides." I referred him to the Clark article on *The Poultry Site,* and he said, "That's what I'm seeing."

Orthopedic (Bone) Disorders

Genetic selection of broiler chickens for rapid growth and gross hypertrophy of the breast muscle has created serious problems of "leg weakness" in the heavy, fastest-growing strains. "Leg weakness" is a euphemism used to describe but not diagnose a long and depressing list of pathological conditions of bones (e.g. tibial dyschondroplasia), joints (e.g. septic arthritis), tendons (e.g. perosis), and skin (e.g. hock burn). . . . Similar problems are seen in the heavy strains of turkeys."

—John Webster, 156

Broiler chickens are not "too mentally unendowed to even stand upright," as a journalist once thoughtlessly quipped about turkeys plagued with similar genetic problems imposed by the poultry industry. They suffer from painful skeletal abnormalities caused by forced rapid growth. Bone calcification cannot keep pace with the rate of growth in these baby birds, a condition that "sometimes results in seepage of pigments from the bone marrow to the surface of the bone when chicken is cooked" ("Dark Chicken Bones").

In 1990, the American Association of Avian Pathologists identified what it considered to be the three most common bone problems associated with the forced rapid growth of present day poultry: angular bone deformities, in which the legs become bowed in or out or twisted; tibial dyschondroplasia, in which the bones develop fractures and fissures; and spondylothesis (kinky back), in which the vertebra become dislocated and/or cartilage proliferates in the lower backbone, pinching on the spinal cord and lower back nerves (Schleifer). A 1996 article in *Feedstuffs* reported that whereas chickens in the early 1960s had an incidence of tibial dyschondroplasia of 1.2 percent, half of them had it in the mid-1990s (Leach).

A 1999 British-Danish study reported nearly 90 percent of broiler chickens with abnormal gaits attributable to genetic manipulation and management practices (Mench 2001). Researchers in the United Kingdom who examined broiler chickens in flocks owned by five major U.K. producers found that over 200 million broiler chickens of the 850 million raised in the U.K. suffered from painful lameness (CIWF 2006b). The Agricultural Research Service of the U.S. Department of Agriculture said in 1998 that "bone weakness" was a common problem associated with the forced rapid growth of broiler chickens and that it was financing research to find ways to strengthen the birds' bones (Gay). This is the same agency that pushes chickens and turkeys to ever greater weights and supports cramming them so tightly in the sheds that they cannot exercise, in order to get more flesh per square foot of floor space and create the type of soft, undeveloped muscles of a baby calf squeezed into a veal crate.

Birds in Chronic Pain

In "Pain in Birds," Michael Gentle writes that the "widespread nature of chronic orthopaedic disease in domestic poultry," added to the fact that there is a "wide variety of receptors in the joint capsule of the chicken," including pain receptors, supports the behavioral evidence that the birds are in chronic pain (242).

Support comes from studies published in the *Veterinary Record* in 2000 in which the analgesic drug carprofen was added to the food of lame broiler chickens. Food that did not contain the drug was also available. The lame chickens quickly identified the food with the pain reliever and chose that food over the other, and their walking ability improved. The researchers concluded that broiler chickens can learn to "self-administer" the analgesic carprofen in their food and that their choosing to do so shows that "lame broiler chickens are in pain and that this pain causes them distress from which they seek relief" (Danbury).

According to John Webster, a professor of animal husbandry at the University of Bristol School of Veterinary Science, while the causes of the painful leg disorders in broiler chickens and turkeys are complex, "most of the conditions can be attributed, in simple terms, to birds that have grown too heavy for their limbs and/or become so distorted in shape as to impose unnatural stresses on their joints" (156). A U.K. study published in 2008 noted that in the past 50 years the growth rate of chickens has increased "by over 300 percent," resulting in "impaired locomotion and poor leg health" (Knowles).

The suffering of these chickens increases dramatically as they become older and heavier, particularly after the fourth week of age. Most chickens are slaughtered between six and seven weeks old, even though a chicken's skeleton is not fully developed even by 10 weeks of age. At six weeks old, only 85 percent of the chick's skeletal frame has been formed, yet that frame is forced to support many times the amount of weight of a normal chicken (Bell and Weaver, 630, 996).

Diseases Traced to Feed Ingredients: Arsenic, Antibiotics, and Manure

Whether it is BSE, nitrofen, dioxin, antibiotics, or other related substances, almost all of the negative headline-making food scandals of recent years had their origin in feed administered to farm animals. . . . [T]he fact that feeds or fertilisers based on chicken droppings are suspected to serve as a medium of bird flu spread or infection makes the extent of the basic problem crystal-clear.

—Pia Mautes, *Meat Processing,* 2006

Farm animal waste is fed to farmed animals. Not only are sick birds shipped directly to consumers; animals who die of undiagnosed diseases before slaughter are fed to chickens and other animals whose body parts, eggs, and milk people consume. Bovine spongiform encephalopathy (BSE, better known as mad cow disease)—the fatal neurological disease caused by feeding ruminant animals such as cows and sheep the central nervous systems of other ruminant animals—has been linked to this practice.

Confronted with the discovery that a variant of mad cow disease can be passed to consumers of beef products in the form known as Creutzfeldt-Jakob disease (Brown 2004), the U.S. and Canada followed the U.K. and other countries in 1997 by banning the feeding of ruminant-derived ingredients—brain, spinal cord, and bones called vertebrae that protect the spinal cord—to ruminant animals destined for human consumption. However, the ground-up brains, bones, eyeballs, spinal cords, stomachs, and intestines of poultry and pigs continue to be fed to cattle, along with the poultry-house bedding, called "litter," into which the birds excrete their droppings, die prematurely, and decompose by the millions each year. Cattle raised next to chicken houses are often grazed on this noxious waste.

Even if feeding ruminant nervous system tissue to other ruminants is officially banned, it is still done, as the U.S. Congress's investigative arm, the General Accounting Office, reported in 2002 (COK 2003). Feeding farmed animals to other animals is integral to the agribusiness economy, where "it's just a matter of using all available protein while reducing waste-disposal problems" (Bueckert).

Why be concerned? In the case of mad cow disease, proteins extracted from sick cows and fed to chickens and pigs contain the same infectious prions that cause mad cow disease, and several of these prions have been located in the skeletal muscle tissue of victims of Creutzfeldt-Jakob disease—the human variant of mad cow disease.

When birds and pigs fed prions are in turn fed to cattle and sheep, the infectious tissue is recycled back to its source. If the danger seems remote, consider that the prion proteins responsible for mad cow disease can withstand the intense heat that is used to render diseased cattle into poultry and pig feed, and that birds have the same type of prion proteins as mammals, including humans. Slaughterhouses slaughter cattle for consumption, regardless of the animal's condition. In 2008, a California company called Hallmark, which supplied ground beef from spent dairy cows to the U.S. Department of Agriculture's National School Lunch Program, was documented by an investigator for The Humane Society of the United States torturing sick cows to force them to stand long enough to reach the kill floor (Kim).

Given the nature of the rendering business, which manufactures pet food and livestock and poultry feed, recycling infective material through farmed animals is inevitable, and animals raised for human consumption are fed the most inferior ingredients of all. Each year, American Proteins, Inc. in Hanceville, Alabama—the largest poultry-product rendering facility in the world—converts 1.9 billion pounds of inedible poultry into 0.6 million pounds of feather meal, fat, and poultry by-product meal for the animal food industries. One thousand truckloads of offal from 20 poultry slaughter plants and dead birds rounded up from 190 farms are brought to the site each week to be converted into two feed grades: "pet-food-grade poultry meals and fat" in one plant and "feed-grade poultry meal, hydrolyzed meal, and fat for the livestock industries" in the other ("American Proteins").

Feed-grade products are a primary source of *Salmonella*, avian influenza, and other diseases in poultry (GRAIN). Poultry litter, which starts out as wood shavings spread on the floors of chicken and turkey houses, is used as a feed ingredient after it has become "nine parts manure" deposited by tens of thousands of birds raised successively in a single house (Morse). Only about every two or three years do U.S. chicken houses get a "total crustout," in which the manure-soaked litter is removed down to the bare floor. Three houses alone fill 35 tractor trailers with 1.4 million pounds of waste (Horton, 79). Poultry litter has been found to be "rich in genes called integrons, which promote the spread and persistence of clusters of varied antibiotic-resistant genes." Researchers at the University of Georgia found that "poultry litter—a ubiquitous part of large broiler operations—harbors a vastly larger number of microbial agents that collect and express resistance genes than was previously known" (Crowe). But that doesn't stop it from being a feed ingredient.

Poultry feed also contains powerful insecticides such as Larvadex, manufactured by the pharmaceutical company Novartis. It includes organic arsenic compounds such as roxarsone, widely used since the 1960s to control internal parasites and promote weight gain and blood vessel growth for heavier, pinker chicken flesh. U.S. chicken producers use a total of 2.2 million pounds of roxarsone each year, more than 95 percent of which is excreted unchanged in chicken waste. The waste in the form of used chicken litter goes into fertilizer, soil, and waterways (Hopey), as well as into poultry feed. A report in *Environmental Science and Technology* states that 20 to 50 metric tons of arsenic end up "unchanged in the roughly 1.5 million metric tons of manure excreted annually" by the 600 million chickens on the Eastern Shore of Delaware, Maryland, and Virginia (Christen). In 1999, the European Union reportedly stopped feeding arsenic to chickens. The U.S. Department of Agriculture reports, "Arsenic concentrations in young chickens are three times greater than in other meat and poultry products." At average levels of chicken consumption—two ounces a day, or the equivalent of one-third to one-half of a boneless chicken breast—"people ingest about 3.6 to 5.2 micrograms of inorganic arsenic, the most toxic form of the element." People who eat more chicken may ingest 10 times that amount of arsenic, which can cause bladder, respiratory, and skin cancers from a daily intake of 10 to 40 micrograms of arsenic (ENS).

Poultry feed also includes fish meal and animal waste that contain harmful levels of toxic peroxides that stimulate excess gastric acid secretions in chickens, resulting in gizzard erosion and lesions of the small intestine (Day). Studies show that feeding chickens poultry by-products probably increases their susceptibility to congestive heart and lung failure through "increased metabolic activity of digestion, absorption, and excretion of protein that cannot be used," along with "poor quality poultry by-product in the ration" (Julian). An example is the disease known as ascites syndrome.

Ascites: Pulmonary Hypertension Syndrome

Ascites syndrome—also known as "waterbelly" and "leaking liver"—is a metabolic disease of the cardiovascular system in the rapidly growing young broiler chickens (Cunningham). It represents the combined effects of poor nutrition, toxic housing, and genetic disabilities. Because of the speed at which the birds are forced to grow, the vascular system "is not as developed as is necessary to support normal oxygenation of blood" (Odum).

The victims are usually found dead on their backs with bloated stomachs, reflecting an accumulation of yellow fluid and clots of material in their body cavities. Birds' lungs do not expand like the lungs of mammals, and the lungs of chickens grow more slowly than the rest of their body. Their lung capacity does not keep pace with the forced rapid growth of their muscle tissue. As a result, there isn't enough capillary space to carry the amount of blood needed to supply the body's oxygen requirements. The effort of the heart to pump enough blood through the lungs results in high blood pressure in the blood vessels of the lungs and in the blood vessels from the right side of the heart to the lungs.

When the blood vessels of the young bird's lungs cannot get enough oxygen, they constrict, decreasing blood flow and increasing blood pressure. To improve the delivery of oxygenated blood to the body tissues, the bird's kidneys produce a hormone that stimulates red blood cell and hemoglobin (the oxygen-carrying protein) production. However, this compensation causes the blood to become more viscous—sticky and adhesive—which in turn forces the right ventricle of the heart to pump even harder to force the more viscous blood into the pathologically constricted blood vessels of the lungs. To adapt to the strain, the bird's heart chambers dilate, and the muscle fibers of the right heart ventricle, which pumps blood returning from the peripheral body tissues back to the lungs for more oxygen, hypertrophy, or thicken.

Together, these events cause the heart valves, which keep the blood flowing in one direction, to weaken, "and the blood begins to leak backwards" (Hoerr). If the bird does not suffocate at this point, "the heart continues to fail, leading eventually to damming up of blood in the veins and the visceral organs" (Odum). As blood fills the veins, and organs swell, the pressure becomes so great that venous blood fluid begins to leak into the organ cavities. The normally low-pressure vessels of the liver are particularly vulnerable. As a result of the now inefficient valves of the heart, blood rising from the liver to the heart begins to seep from the surface of the liver until the ability of the abdominal membrane to reabsorb it is surpassed and the abdominal cavity fills with fluid.

I watched a similar disease develop in our rooster Phoenix, whom we rescued from a Tyson facility in Maryland. Already, as a tiny chick, Phoenix had ominous sounds in his chest. Eventually, the right side of his face filled with fluid and his right eye swelled shut. His crow gurgled as if he were under water. At the least stress his comb turned blue from lack of oxygen. He collapsed and died in the yard at 14 months old.

Ascites is often underway even before the birds hatch because of "industry demand for increased incubator egg density and chick output, producing mild-to-severe embryonic hypoxia" (Odum). Since most hatcheries are "not designed to have a 100 percent hatch because of insufficient oxygen available in the machines," many chicks break out of their shells already coping with cardiopulmonary disease (Bell and Weaver, 698).

Inside the Chicken House

The little chickens take up very little room in the huge houses, but as they grow the floor gets crowded and the birds are lucky to stay healthy in that environment.

—Mary Clouse, former chicken grower in North Carolina (2003)

The dim light is typical of commercial husbandry, while bright light is consistent with that preferred by broiler chickens given a free choice test. . . . In dim light, the exploratory motivation is diminished.

—C. M. Wathes et al. 2002, 1609

"a sea of stationary grey objects"

—Andrew Purvis, *The Guardian*

From the hatchery the chicks are taken to live in an oxygen-deficient shed filled with pathogenic microbes, carbon dioxide, methane gases, hydrogen sulfide, nitrous oxide, excretory ammonia fumes, lung-destroying dust, insecticide sprays, and dander (tiny particles of feathers and skin) suspended in the air and embedded in the litter.

In an article in the *New Yorker*, Michael Specter described his visit to the Eastern Shore of the United States to investigate broiler chicken farms. Here is what he found.

> Except for the low hum of a ventilation system, the sheds that I approached were quiet. Every window was covered with thick blackout curtains, and it seemed as if nothing at all were inside. After a few stops without finding a farmer at home, I decided to try one of the doors. It wasn't locked, so I unfastened the latch, swung it open, and walked inside. I was almost knocked to the ground by the overpowering smell of feces and ammonia. My eyes burned and so did my lungs, and I could neither see nor breathe. I put my arm across my mouth and immediately moved back toward the door, where I saw a dimmer switch. I turned it up.
>
> There must have been 30,000 chickens sitting silently on the floor in front of me. They didn't move, didn't cluck. They were almost like statues of chickens, living in nearly total darkness, and they would spend every minute of their six-week lives that way. Despite the ventilation system, there wasn't much air in the room, and I fled quickly (Specter, 63).

Tunnel Ventilation Housing

These tunnel houses, as told us by a man who currently has them, have only the dimmest of light. He said his birds at five weeks can hardly stand because their legs are so weak, and with no natural light or exercise, their joints are too soft to carry the weight. Simmons is forcing ALL their growers to go to this type of housing.

—A neighbor who requested anonymity regarding
Simmons Foods in Arkansas in 2003

The latest proposal by Rural Funds Management is to build a couple of MEGA broiler complexes near Melbourne [Australia] with 64,000 birds in EACH shed. These are the new "state of the art" tunnel ventilation sheds, which, they contend, you can cram more birds in due to the ventilation system.

—Patty Mark, 2006

The chicken-house environment that Michael Specter of the *New Yorker* fled from is called tunnel ventilation, or tunnel housing. Whereas industrialized chicken houses used to have heavy curtains that could be raised and lowered on the long sides of the 400- to 500-foot-long buildings, allowing some fresh air and natural sunlight in, tunnel ventilation houses are built with solid walls or with the side curtains tightly attached to the house frame in order to "black out" the house inside. Tunnel ventilation housing requires the continuous running of huge computerized fans to pump air out of the chicken houses through air inlets that are covered with evaporative cooling pads in hot weather. However, with houses 400 to 600 feet long and 40 to 45 feet wide, "even the most powerful exhaust fans leave 'dead spots' in corners, near ceilings, along the walls, and in the middle where the air is not pulled evenly," says former North Carolina chicken grower Mary Clouse.

The dim interior of these tunnel houses is designed to keep the birds quiet and reduce their movements to getting up only to eat, drink, and sit down again, so they'll gain weight more quickly. Whatever the "benefits" of this system may be, Clouse says that "the dim light bothers farmers who are accustomed to working out of doors in bright sunlight, and it also seems unnatural to raise any living animal in the dark all the time. It is difficult to see the birds, and sudden light or a flashlight frightens the birds into piling up, causing injuries and suffocation."

There is also machinery malfunction to worry about. According to Clouse, growers are afraid to leave their farms for more than a few minutes. "Computer-generated alarms attached to a beeper on their belt, beside their bed, or in the house alerts them that the temperature is too high or too low in a certain house, water pressure is down indicating a leak somewhere, the fans are off, or the power has cut off in one of the houses. Typically they must race to the chicken houses to locate the problem and fix it quickly or they stand to lose all 25,000 birds in any one house in a matter of minutes from suffocation or dehydration."

Toxic Waste Environment: Poisonous Gases and Disease-Causing Microbes

Numerous surveys of air hygiene in broiler [chicken] houses have revealed a dense miasma of aerial pollutants.

—C. M. Wathes, "Air Hygiene for Broiler Chickens," 2003

Birds suffocate even when the computers are working. The combination of toxic gases, airborne microbes, insecticides, organic and inorganic dust, dense crowding, and respiratory diseases make suffocation a daily event in the chicken houses. Poultry house air is polluted with many gaseous substances, but excretory ammonia is the most prevalent, along with nitrous oxide, carbon dioxide, hydrogen sulfide, and methane (Kristensen and Wathes, 235–236). Eight hours of the standard minimum amount of ammonia in the average commercial chicken house—25 to 35 parts per million as established by the U.S. Office of Safety and Health Administration (OSHA)—is considered the maximum allowable concentration for an adult human being (Carlile, 99; Malone).

This is an important fact given that chickens need three times more air volume than humans per kilogram of body weight to meet their oxygen requirements (Carr and Nicholson). A person entering a chicken house holding thousands of birds experiences the burning sensation in the eyes and deep in the throat described by Michael Specter in the *New Yorker*. What causes this?

Excretory Ammonia

Ammonia is water soluble and can thus be absorbed in dust particles and litter as well as in mucous membranes. It is toxic to animal cells.

—Kristensen and C. M. Wathes, 237

The detrimental effects of ammonia in the poultry house have been known for many years.

—Mac Terzich, DVM (1995)

Excretory ammonia is a colorless irritant gas produced by microbial activity on the nitrogen excretion content, uric acid, in poultry manure (Carlile, 99). Though not a problem under natural conditions, in the densely packed poultry unit, the breakdown of manure becomes poisonous. Poultry workers experience eye, lung, and nasal irritation. They develop headaches, nausea, wheezing, coughing, phlegm, and other respiratory symptoms. Prolonged intermittent exposure can lead to chronic respiratory disease and to feeling unwell much of the time. This situation is bad for people, but it is worse for the chickens, who cannot escape their noxious surroundings.

Chickens, turkeys, and pigs can detect and will choose to avoid ammonia concentrations at or below the recommended exposure limit of 25 ppm (parts per million). In one study, laying hens given a choice between fresh air, 25 ppm of ammonia, or 45 ppm of ammonia for a period of six days "significantly preferred fresh air to the ammoniated atmospheres and were found to forage, rest, and preen significantly more in fresh air than in ammonia" (Kristensen and Wathes, 241). Studies in which broiler chickens were made to choose between being active in a well-lit environment full of ammonia fumes and sitting quietly in a dimly lit environment filled with fresh air chose the "comfort of fresh air in which to rest" (Wathes 2002, 1609).

No wonder given that ammonia concentrations of just 5 ppm (undetectable by the human nose) have been found to "irritate and injure the protective lining of the chick's respiratory system, causing increased susceptibility to respiratory disease" (Bell and Weaver, 837). Ammonia dissolves in the liquid on the chicken's mucous membranes and eyes to produce ammonium hydroxide, an irritating, alkali-causing ammonia burn that stimulates the production of excessive mucous in the trachea. This mucous mats, and ultimately destroys, the tracheal cilia, which serve to block the entry of harmful agents into the system, inviting colonization of the airways by airborne microorganisms such as *E. coli* bacteria, Newcastle disease virus, and avian influenza virus. Chickens exposed to 20 parts per million of ammonia for 42 days develop pulmonary congestion, swelling, and hemorrhage. Increased ammonia thickens the arterial walls and shrinks the air capillaries in exposed birds. Ammonia stress in young chickens and turkeys harms their developing immune systems, causing "severe vaccine reactions" (Carlile, 101). "Ammonia in the air is absorbed into the blood of turkeys and causes immunosuppression," according to the National Turkey Federation (NTF).

Chickens exposed to 60 ppm of ammonia—a common level in broiler chicken houses—develop keratoconjunctivitis, a painful inflammation and erosion of the eye cornea and the conjunctiva, which is the mucous membrane lining the inner surface of the eyelids and covering the front part of the eyeball. Afflicted birds cry out in pain. They rub their head and eyelids against their wings and may not eat. High levels of ammonia in the poultry facility cause a cloudy appearance in birds' eyes and can lead to blindness. Birds blinded by ammonia die of hunger and thirst, unable to find food and water (Kristensen and Wathes, 239–240).

Ammonia is an economic problem, because it increases bird mortality, retards birds' growth rate, and impairs the birds' immune systems, thus increasing their susceptibility to *E. coli* infections, leading to condemnation and downgrading of carcasses at the slaughter plant. New technologies to reduce ammonia and dust levels in chicken and turkey houses are often touted but don't seem to work. Growers are urged to note when the ammonia concentration in the poultry house exceeds 25 ppm, an innately harmful level. They are reminded to check concentrations at bird level, close to the litter, where the uric acid decomposes and the combination of water vapor, litter moisture, heat, carbon dioxide, micropathogenic activity, and ammonia is most intense. However, their noses quickly acclimate, particularly as their own respiratory problems increase. People who have worked in poultry houses for years "often cannot detect levels below 50 ppm" (Malone).

Ammonia levels are especially high in poorly ventilated houses and during the winter, when ventilation is reduced to conserve heat. At such times, the ammonia concentration can go as high as 200 ppm (Carlile, 1). Condensation during the winter wets the litter, releasing ammonia fumes into the air and increasing painful breast blisters and manure burns in the birds. A poultry grower recalls the human experience:

> During the winter, when vents could only be cracked because of the frigid outside air, we were often forced from the building, gasping from the high concentration of ammonia. Breathing becomes painful if not impossible; eyes sting and water. During such times we toyed with the idea of liberating the birds, and ourselves, from the confines of the windowless, stinking imprisonment (Baskin, 38).

Overcrowding

Poultry [are] raised in a commercial broiler house using high-density rearing—where up to 50,000 birds may be raised in a single chicken house.

—TV, "Less Stressed Chickens Mean More Dollars for Poultry Growers," *College Park: The University of Maryland Magazine*, 2002

So tightly packed are the birds, they sit immobile in their own soiled litter, with blistered breasts and ulcerated wounds from the ammonia in their excreta.

—Andrew Purvis, *The Guardian*, September 24, 2006

To the toxic air is added the lack of living space. A Perdue pamphlet asserts, "Chickens are naturally a flocking animal, so the question of the space they need is irrelevant" (Curtis). Consider the following proposition: "Humans are gregarious by nature, so the question of the space they need is irrelevant." Does the human instinct to "flock" together and be social mean that as individuals we have no spatial requirements? The need for personal space is as basic to birds and mammals as is the need to be together. (This is obvious, but for a scientific discussion, see "Social Behavior Effects on Spacing" in James V. Craig's book, *Domestic Animal Behavior.*)

Personal space is completely relevant to a chicken's well-being. As Smith and Daniel observe in *The Chicken Book*, "Chickens, like all living creatures, love to be free." If anything, it is people, not chickens, who would often rather be pent up than free (Smith and Daniel, 337).

The National Chicken Council tells people that "the physical welfare of animals is very important to the broiler chicken industry" (Lobb). In reality, crowding the birds causes more diseases and deaths, but the tradeoff is accepted because more pounds of flesh are obtained as a result of the volume of birds being raised. The chicken industry tells the public that economic profitability cannot be achieved without careful attention to the welfare of the chicken, but this is not how the system actually works. Chickens can be profoundly mistreated and still "produce," just as profoundly mistreated humans can be overweight, sexually active, and able to produce offspring. Like humans, chickens can "adapt," up to a point, to living in slum conditions. Is this an argument for slums?

The welfare of the birds and the economics of floor space per bird in the broiler chicken house are inversely related. North and Bell explain: "The more you crowd broilers and roasters, the poorer the results. However, as floor space is reduced per bird, the greater the weight of broilers produced in the house, and this will, up to a certain point, increase the return on investment. . . . [L]imiting the floor space gives poorer results on a bird basis, yet the question has always been and continues to be: What is the least amount of floor space necessary per bird to produce the greatest return on investment." For example, reducing floor space increases mortality and breast blisters, but it also increases "the pounds of broilers raised in a given house during a 12-month period" (456–457). Instead of counting the number of birds per floor space, industry says it is "more appropriate to define density in terms of pounds or kilograms of market liveweight per square foot or square meter" (Bell and Weaver, 854–855).

By reducing the birds' living space from a square foot to a half square foot per bird, twice as many birds die. However, almost twice as many birds survive long enough to go to slaughter. As a result, the producer gets seven and a half pounds of flesh per square foot instead of just four or five—almost twice as much flesh per square foot of floor space. A house designed for 27,500 birds gives to each four-and-a-half-pound chicken "0.8 square foot of floor space, or 13.4 birds per square meter" (Bell and Weaver, 822–823).

Recall that one square foot equals 144 square inches and half a square foot equals 72 square inches. Compare this to the fact that a three- to four-pound chicken needs a minimum of 74 square inches merely to stand, 197 square inches to flap wings, 135 square inches to ruffle feathers, 172 square inches to preen, and 133 square inches to scratch the ground" ("British Scientists").

These are basic biological activities. Crowding enforces inactivity, reducing the energy—food—that would otherwise be "wasted" by a normally active bird instead of being converted to the flaccid flesh of a sedentary inmate. Crowding encourages passive adaptation to a deadening environment. Chronic deterioration of lively alertness to lethargy in intensively confined chickens is misrepresented as proof that the chickens are "happy," or, alternatively, that they are brainless and unresponsive.

Parent Flocks: Blackouts and Food Restriction

Broiler breeders—the parent stock who produce broilers—have the same huge appetites as their progeny and have to be maintained on very severe food restriction so that they are able to reproduce.

—Ian J.H. Duncan, "Animal Welfare Issues in the Poultry Industry: Is There a Lesson to Be Learned?" (2001)

To prevent injury to the backs of the females during mating, the toes of day-old meat-type cockerel chicks should be trimmed at the hatchery. Trimming should occur at the outer first joint of the back toe of each foot. . . . In some cases, it may also be necessary to trim the two inside spurs. Use an electric beak trimmer or toe clipper to remove both toes and spurs. This is a very delicate procedure. . . . To reduce comb injuries, and to help identify sexing errors in the female line, the combs of male chicks are usually trimmed at the hatchery. A disadvantage of this practice is that the exclusion of males from the female feeder during production is made more difficult when a grill is used for sex-separate feeding.

—Bell and Weaver, 625

Chickens raised to produce the eggs that become "broiler" chickens are called broiler breeders. Male and female chicks are raised separately until they are about five months old. At that time, the young hens and roosters are brought together in laying houses that hold from 8,000 to 10,000 birds, with 10 to 12 roosters for each 100 hens. The flock is maintained for breeding purposes for about 10 months. At a little over a year old, they are sent to slaughter. Their eggs are gathered and taken away to the mechanical hatcheries; the parents never see their chicks. Throughout the breeding period, roosters are rigorously culled (removed and killed) for infirmity and infertility and because "if a particular male becomes unable to mate, his matching females will not accept another male until he is removed" (Bell and Weaver, 641).

Breeding flocks are kept on floor systems that are part litter (wood shavings) and part wooden or plastic slats. Each adult bird gets two square feet of living space (Bell and Weaver, 821). Breeder houses are equipped with mechanical nest boxes—one for every four hens—and feeders that are similar for the roosters and hens except that "the wires that allow the birds to slip their heads in to reach the feed are closer together on hen feeders than on rooster feeders" ("Growing Up with Tyson").

The reason for this difference is to keep the roosters from eating the hens' food as they would otherwise do. Breeding hens and roosters are kept on semi-starvation diets because the characteristics bred into broiler chickens to produce rapid and excessive weight gain in the chicks cause obesity, infertility, and mating problems in their parents (Leeson, 49–50).

"Multitudes of problems plague overweight broiler breeder hens," states an article in *Feedstuffs* (Muirhead 1993b). These hens suffer from complications including malfunctioning ovaries and breathing problems. They develop congestive heart and lung failure. Our hen Olivia became so fat, even with exercise, that her abdominal air sacs were pinched off (as shown in the X-rays), and she nearly died, barely able to breathe, at 10 months old.

The breeding hen's malfunctioning ovaries result in erratic laying patterns, soft eggshells, low and short-lived fertility, double and triple yolks being laid, embryo loss, and so on. Left to eat as they please, the roosters become so large, unwieldy, and disabled that they cannot mate properly or even move without pain. Even as babies, their legs tremble when they try to stand up and walk a little, like the legs of very old men. To curb these effects, broiler breeder chickens are kept in semi-darkness in "blackout houses," which, added to the ammonia fumes, can create eye disorders, including blindness. In addition, as has been noted, they are kept on semi-starvation diets designed to control their weight and reduce their food intake ("Chicken-Rage"). Typically, a whole day's food is withheld from the birds every other day starting at three weeks old, or they are fed very small portions in "every-day feeding since it is more efficient" (Leeson, 49). The chickens rush pitifully to the feeders when the food is restored, often injuring their feet and other parts of their bodies in their desperation to eat. Bacteria invade the tissues and bloodstream following these injuries to the skin, particularly the feet, which become inflamed with a painful footpad disease called bumblefoot.

Feed-restricted chickens gorge themselves when the feeders are refilled, enlarging the capacity of their crop and gizzard to hold even more food, adding to the birds' frustration. On days when food is withheld, they peck compulsively at spots on the floor, at the air, nonfood objects, and each others' heads, and, if permitted, they drink up to 25 percent more water than normal to compensate for the feeling of emptiness. Because this results in loose droppings and wet, ammoniated litter, managers are urged to restrict the availability of water to "birds looking for feed" (Bell and Weaver, 631–633).

Nozbonz

In addition to separate feeders for the roosters and hens, many broiler breeder roosters are implanted with a nasal devise called a Nozbonz to prevent them from poking their heads through the hens' food restriction grill. Maybe 50 percent of producers in North America use them, according to a University of Georgia researcher in 1999 (Mauldin). They're used on the Eastern Shore in the U.S. where I've seen roosters with the Nozbonz fused into their faces, sticking out on either side. The Nozbonz is a two-and-a-half-inch plastic stick that is jammed through the bird's cere (nasal cartilage) when he is five months old. Suzanne Millman, a researcher at the University of Guelph in Ontario, said she "did a few of the birds myself" and found that "it definitely takes quite a bit of pressure to get the Nozbonz through the septum."

As bad as all this is, Millman writes, "Nozbonz are not effective for the purported objective." Since the roosters are "extremely food motivated," they "soon try their own experiments to access the female feeders. Females have a much higher feed allotment than the males due to demands from egg production, so males finish their meals in about 15 to 20 minutes and females take several hours. Males will turn their heads to get into the grills and become vulnerable. If attacked by other birds, or frightened, they then often try to pull back and panic. They may injure or even kill themselves. They also sometimes rip the Nozbonz out in the struggle."

Abnormally Aggressive Roosters and Hens

A new problem emerged in the poultry industry in the 1990s. An increasing number of reports described broiler breeder males being very aggressive toward females. This is highly unusual because male domestic fowl dominate females passively and seldom show any overt aggression toward them.

—Ian Duncan, "Animal Welfare Issues in the Poultry Industry: Is There a Lesson to Be Learned?" 2001, 213

The Nozbonz experiment was part of an investigation into the causes of unusual aggression in broiler breeder chickens, a new manmade pathology variously attributed to the birds' impoverished environment, food frustration, and genetic malfunction (Duncan 1999). In one study, the introduction of "bales of plastic-wrapped wood shavings" to broiler breeder hens was said to "dramatically" reduce their unnaturally aggressive behavior ("chicken-rage"). Studies by Millman and Duncan led them to speculate that attacks by broiler breeder roosters on hens is a genetically based courtship disorder, since even roosters who have been bred for cockfighting "show little, if any, aggression toward females" (Duncan 2001, 213).

The courtship theory recurs in Temple Grandin's book *Animals in Translation,* which proposes that genetic manipulation of chickens for abnormally fast growth has somehow eliminated the rooster's courtship dance around the hen that tells her to crouch into a sexually receptive position. When the hen fails to crouch or tries to escape from being mounted by the rooster, according to the theory (which Grandin mistakenly presents as a fact: "She doesn't crouch down unless she sees the dance. That's the way her brain is wired."), the rooster attacks her with his spurs or toes and slashes her to death, an example of what Grandin calls "warped evolution" in animals bred for single traits at the expense of overall fitness (Grandin 2005b, 69–72).

The trouble with the courtship theory is that these kinds of hens do crouch, even without a rooster. If you so much as lay your hand gently on their backs or approach them from behind, they will walk or run a little, then stop and crouch abruptly during the spring and summer mating season. In addition, a rooster's courtship dance around a hen does not automatically cause her to crouch. She may simply curve around and away from the rooster, signaling the end of the encounter. She may run away or continue her dustbath or whatever else she was doing when he approached. Her response does not provoke the slashing attack described by observers of broiler breeder chickens in laboratories and commercial breeding operations.

Part of what is wrong with these birds is that they have been artificially bred to become sexually mature at around three months old instead of the normal six months, so that halfway out of their infancy, they have adult sex hormones driving them, without the neurobiological maturity of an adult bird. Add to this the barren environment, eye-impairing darkness, and chronic hunger, plus the fact that the broiler breeder rooster's body, legs, and feet are too big for the hens who are themselves abnormally heavy, disproportioned, and slow moving, and have thin, easily torn skin, and nowhere to escape to, and you have predictably abnormal behavior. "Spent" broiler breeder hens that we've adopted into our sanctuary arrive in terrible condition, with large patches of raw bare skin and ragged feathers. Even the soft tuft of feathers that nature designed to hide and protect their ears is missing, exposing the ear hole. This is something you simply don't see in young adult birds.

"Liquidation"

After 40 or so weeks of producing fertile eggs and being plagued by hunger, mutilations, toxic ammonia, fear, pain, stress, and disease, broiler breeder chickens are "liquidated" and turned into human animal "food" and nonhuman animal "feed" and pet food products (North and Bell, 404).

Broiler Chickens in Cages and Elevated Floor Systems

Attempts have been made to raise broiler chickens and their parents (separately, that is) in multi-tiered cage systems in order to fit more birds into a building (Dudley-Cash 1995, 19). So far, leg problems, breast blisters, poor growth, and labor costs have been cited as reasons against these systems. In the mid-1990s, plastic mesh floors with automated manure- and bird-removal systems were said to have solved problems, but whereas multi-tiered cage systems have been developed in North America, Eastern Europe, and the Middle East, they are not used by the U.S. industry, which has invested instead in the tunnel ventilation housing with litter floors that was discussed earlier in this chapter.

Because the huge chicken flocks in the U.S. are raised consecutively on the same bedding, and replacing this soiled litter for each new flock is considered "not economically justified under most circumstances," researchers at the University of Maryland Eastern Shore are trying to develop a type of elevated, plastic-mesh floor and ventilation system that would eliminate the use of litter altogether. "Futuristic as it may seem to the U.S. industry," stated an article in *WATT Poultry USA* in 2007, "rearing birds on an elevated flooring system, not particularly in cages, is being evaluated at locations outside of the USA." Companies may one day have to face the disappearance of bedding material from the market (Blake, 41).

FEATHERLESS CHICKENS

It's a prime example of sick science and the suggestion that it would be an improvement for developing countries is obscene.

—Joyce D'Silva of Compassion in World Farming
quoted in *New Scientist* (Young 2002)

Still another research enterprise is the development of featherless chickens for industrial farming in the Middle East (van der Sluis 2007). A professor at the Hebrew University of Jerusalem's Department of Agriculture in Israel has made a career of this effort, and the U.S. Department of Agriculture's Agricultural Research Service breeds and slaughters featherless chickens to study such things as whether featherlessness reduces *Campylobacter, E. coli,* and other bacteria on slaughtered chickens' skin. It doesn't (Durham 2002).

Proponents claim featherless chickens would fare better in hot climate zones, but critics say featherless chickens "suffer more than normal birds." Roosters "have been unable to mate, because they cannot flap their wings," and featherless chickens of both sexes are "more susceptible to parasites, mosquito attacks, and sunburn" and are "very susceptible to any temperature variations—especially as young birds" (Young 2002).

In the U.S., normally feathered chickens with bare skin areas have been shown to have "lacerated backs and flanks," especially in the "slow-feathering" male birds and in birds grown to weights above six pounds. According to the U.S. Department of Agriculture, infections resulting from these types of injuries are an increasing source of slaughterhouse condemnations in the United States (Shane 2007).

Dead Bird Disposal

If you have 60,000 in two houses with a 5 percent death loss, then you have 3,000 [chickens] to dispose of.

—"Deep Throat Chicken" 1994

Matthew Rieley, three, is learning his numbers the way all eleven sisters and brothers did—by counting the dead chickens the children pick up and compost daily.

—Tom Horton, "42-Day Wonders," 2006

A problem with dead bird composting is that it allows nitrate residue to go back to the land.

—Sharon Morgan, "Chicken Disposal Idea Wins National Award," *The Delmarva Farmer* 1991

Millions of chickens die in the sheds each year of heat suffocation, medication reactions, and diseases before going to slaughter. Their bloated, decomposing bodies and skeletal remains can be seen on the poultry house floor, trashed in cans inside and outside the chicken houses, and dumped on the ground just outside the door or around back of the building. Eventually, the accumulated carcasses are buried in the ground, burned, dropped down feed shoots, composted, or picked up by rendering companies to be turned into feed ingredients. Air and ground water pollution, insect infestation, rodents, and odors result from this mortality.

Small, sick, injured, and deformed birds are "culled"—weeded out and killed. Feeders are raised so that slow-growing birds starve to death. Growers walk through the chicken house casually breaking the necks of "cull" chicks and toss them, writhing, on the floor against the wall, as is shown in *Commercial Chicken Production*, a video produced by the University of Delaware. Chickens dead or alive are dumped in landfills to be bulldozed or otherwise buried along with the trash. A Tyson grower in Alabama said she hit sick and injured birds on the head with a stick and took live birds to the landfill (Farm Sanctuary 1993). Others report stuffing sick and slow-growing birds into large Tyson freezers to die alongside the already dead birds waiting to be picked by the rendering company ("Humane Society").

In the *New Yorker*, Michael Specter described watching a grower dump a live, six-week-old chicken with crushed bones and a mangled head into a dumpster along with the dirt load. The chicken's "vastly oversized chest was heaving up and down, and its beak dug slowly into the dirt" (63).

Mass Extermination and Disposal of Bodies

Whatever the shortcomings of current techniques, federal and industry officials insist they're ready for a big outbreak [of avian influenza in poultry flocks]. But they acknowledge they'd probably use [firefighting] foam and gas as needed—and some uglier methods too.

—Jeff Donn, Associated Press, 2007

After trying to burn the bodies (too expensive), burying them (an environmental hazard), and trucking them to rendering plants (risking further spread of the disease), poultry experts believe the safest means of disposal might be to roll the dead birds into a small hill and let the blistering heat of decomposition burn away the disease inside.

—William Wan, "In Pile of Waste," 2006

In addition to the daily mortalities, entire populations of birds are destroyed and disposed of by contract crews hired to respond to power failures, bird oversupplies, and the avian influenza viruses that rage periodically among the fragile poultry flocks (Bell and Weaver, 162). For example, in 2002, the Commonwealth of Virginia destroyed 4.7 million turkeys and chickens on 197 farms to combat a strain of avian influenza virus. In 2003, the U.S. Department of Agriculture destroyed more than 3.5 million chickens in California to halt the spread of the exotic Newcastle disease virus. In 2004, the Canadian government destroyed 19 million birds to curb an avian influenza outbreak in British Columbia. And in 2007, the British government destroyed 159,000 turkeys on an avian influenza-infected farm in East Anglia, the heart of Britain's poultry industry (UPC 2007).

There are no welfare regulations for these mass killings. The International Organization for Animal Health—the OIE—has produced "Guidelines for the Humane Killing of Animals for Disease Control Purposes," but, typical of such documents, the guidelines are not truly humane or of any real practical use (OIE). They merely codify standards accepted by industrialized nations as "humane" in accordance with standard business practices. Sherri Tenpenny writes the following in *FOWL! Bird Flu: It's Not What You Think*:

> While international bodies call for the massacre of all birds exposed to highly pathogenic influenza viruses, there are no international regulations governing how the birds are to be killed. Methods include burning, drowning, gassing, and live burial. Throughout the world, tens of millions of birds have been brutally killed. In Bali, 228,000 chickens were reportedly either burned alive or kicked and beaten to death. In Thailand, more than 10 million birds were destroyed, most by burying them alive. "As soon as the first 500 died, we had to bury the other 20,000 alive," a Thai farmer told a reporter (Tenpenny, 146).

These "depopulations" are largely or entirely funded by tax dollars. In 2003, the U.S. Department of Agriculture provided $100 million to exterminate more than 3 million birds and compensate owners during the viral exotic Newcastle disease outbreak in California (Olejnik). In 2005, the U.S. poultry and egg industries and USDA proposed "100 percent" reimbursement to producers enrolled in the National Poultry Improvement Plan for "the cost of purchase, destruction, [and] disposal" of poultry infected with or exposed to low pathogenic avian flu viruses, along with the cost of "cleaning and disinfection of contaminated barns and equipment and 100 percent of the cost of any surveillance, vaccination, monitoring, and euthanasia *[sic]* associated with an eradication effort." Producers not enrolled in the NPIP plan would still receive a 25 percent indemnity payment (Rhorer).

Media Management of Poultry Massacres

In 2007, researchers from the University of Delaware destroyed 25,000 male turkeys using firefighting foam on a farm infected with avian influenza in West Virginia. The birds' horrific death can be glimpsed in an article in *Lancaster Farming* (PA), which describes how the killing crew "struggled with their equipment," faced "foam quality issues, pump failure, and worker fatigue," and buried the live turkeys under various brands of foam including foam that was "20 years old and had 'sludge' in the bottom of the container." The head researcher said he was "pleased with the results," which he called a "success" (Espenshade).

WATT Poultry USA reported that the turkeys were "all euthanized" with the foam (Thornton). *West Virginia Media* reported a heroic destruction of "nearly 1 million pounds of the birds" to protect the food supply (Darst), and the Associated Press said merely that the turkeys were "dispatched" with the foam (Donn). Producers have been warned by industry officials to conduct their massacres out of sight, "so as not to unduly distress onlookers" (Galvin).

Consumers and Animal Welfare

FMI [Food Marketing Institute] learned that consumers were concerned about the animals, but they also wanted to continue to purchase foods of animal origin. "They also told us they didn't want to know all the details about how animals are raised and processed into food."

—Sally Schuff, "Welfare Best Practices Emerge," 2007

Throughout the world, chickens are raised in the manner described in this chapter, except that in countries like China, Thailand, Mexico, and Brazil, the situation is even worse—filthier living conditions, more brutal treatment, no "welfare" aspect.

Responding to welfarist pressure, the National Chicken Council, which at first refused to provide the Food Marketing Institute and the National Council of Chain Restaurants with welfare guidelines as a basis for the new auditing systems being put in place, finally agreed to do so. It was easy. The *National Chicken Council Animal Welfare Guidelines and Audit Checklist as of April 5, 2005* merely outlines commercial practices that are already being done. Atmospheric ammonia in the chicken houses "should not exceed 25 parts per million" (a highly toxic level). Birds weighing "4.5 pounds liveweight" may be stocked at "6.5 pounds per square foot." Birds weighing more than 5.5 pounds liveweight may be stocked at 8.5 pounds per square foot, and so on.

Following suit, Tyson Foods issued a press release on November 21, 2003, announcing the establishment of an Office of Animal Well-Being in Springdale, Arkansas, to "ensure compliance with the National Chicken Council's Animal Welfare Guidelines and the Recommended Animal Handling Guidelines formulated by the American Meat Institute."

In 2007, Temple Grandin reported that while she believes audits have brought about "huge improvements" in poultry and livestock facilities—the "good" poultry plants are doing a "great job, despite using the same equipment as the bad plants"—there are still "some very bad plants operating outside the auditing system." Problems include "high death loss, rough catching, broken cages, and failure of bleeding machines," along with mistreatment of "half-hatched or weak baby chicks" and overall "poor management" (Grandin 2007).

On May 7, 2007, the Council of the European Union, under welfare pressure, issued a press release announcing a Council Directive "on the protection of chickens kept for meat conditions." It allows atmospheric ammonia fumes in concentrations of 20 parts per million and bird-stocking densities of 6.76 pounds (33 kg/m^2) per square foot of floor space (33kg/m^2) and 7.99 pounds per square foot of floor space (39 kg/m^2). In other words, like its U.S. counterpart, it codifies business as usual.

Eating Misery

Some critics have argued that when we eat the flesh and eggs of creatures who have been treated with such complete contempt, we assimilate something of their experience and carry it forward into our own lives. In *Diet for a New America*, John Robbins asks us to consider the consequences of eating the results of such misery. Could it be, he asks, that when we consume the flesh and eggs of these poor birds, "something of the sickness, misery, and terror of their lives enters us? Could it be that when we take their flesh or eggs into our bodies, we take in as well something of the kinds of lives they have been forced to endure?" (Robbins, 69).

Understandably, one does not like to think that the dead bird one is about to consume embodies the misery and cruelty endured by the bird when alive. So far, all that society has required is that the events that produce the carcass be removed from consciousness. The possibility that the individual's suffering could somehow persist and be present in the body tissues and "juices" about to be ingested is frightful. But is it fanciful?

Consumers and Contamination

Disease-causing organisms are ubiquitous in poultry-producing facilities all around the world.

—Mojtaba Yegani, *World Poultry*, January 16, 2007

We all know that pathogens of all forms, such as bacteria, fungi, and viruses, are everywhere in the animal-production environment and will remain, regardless of techniques adopted.

—John P. Blake, *WATT Poultry USA*, May 2007

Poultry is the most common cause of food poisoning in the home.

—Michael Greger, *Bird Flu: A Virus of Our Own Hatching*, 47

Once bacteria and other microbes were just a "theory." We could not see them, yet they existed. Historically, the U.S. federal Meat and Poultry Inspection Acts do not mandate inspection for disease microbes in animals slaughtered for food. However, contamination of poultry products with poisonous bacteria is not simply the result of an inadequate inspection system.

The destruction of the family life of the chicken is a major primary cause. Mass-produced chickens and turkeys are unable to obtain the parental protection they need. Before the advent of large-scale production systems, normal intestinal microflora, a mixture of hundreds of different types of harmless bacteria that occupy the lining of each chicken's and turkey's intestinal tract to provide immunity, were transferred from adult birds to their offspring by way of their droppings, creating an immediate natural defense. As the chick grew, its own microflora developed, so that by six months old the bird was healthy and strong (Fowler).

In modern production, not only is the intestinal microflora of the young chicken incomplete; it is disrupted by the use of antibiotics. Antibiotics pump up the birds artificially by causing water retention and by disturbing the composition and interaction of the birds' microflora, thus increasing susceptibility to colonization by *Salmonellae* and other harmful bacteria.

On July 28, 2005, the U.S. Food and Drug Administration ordered the poultry industry to stop using the antibiotic Baytril based on evidence that its use was preventing its human-label counterpart, Cipro, from treating people sick with *Campylobacter* infections, often the result of eating contaminated chicken and turkey products (Kaufman 2005). The poultry industry has countered that limiting the use of antibiotics in birds raised for food may not reduce, and could even increase, *Salmonella* and *Campylobacter* contaminations, adding to the human health risk (Lundeen).

Recent reports suggest that *Salmonella* and *Campylobacter* infections are rising. In 2007, a meeting of microbiologists in Toronto observed that the role of "bacteria normally found in the gut, especially food-borne microbes, in transmitting resistance genes, is becoming a concern to the scientific community" (Merrett).

Campylobacteriosis—which causes severe abdominal cramps, nausea, and diarrhea, and can cause a paralytic disease in people with fatal nerve damage known as Guillain-Barre syndrome—has "trebled in the past 15 years" in New Zealand, where retail chicken products and packaging have been found "literally dripping with campylobacter" (Watt). Researchers in the European Union call salmonellosis and campylobacteriosis "by far the most frequently reported food-borne disease in the EU" ("Study Finds Salmonella").

In 2007, *Consumer Reports* announced that tests on 525 chickens purchased from U.S. supermarkets and specialty stores in 23 states showed 84 percent of chickens contaminated with *Campylobacter* and *Salmonella* bacteria—a substantial increase over 2003 tests showing 49 percent of chickens infected. In addition, 84 percent of the *Salmonella* and 67 percent of the *Campylobacter* bacteria showed resistance to antibiotics. "When we took bacteria samples from contaminated broilers [chickens] and tested for sensitivity to antibiotics, there was evidence of resistance not just to individual drugs but to multiple classes of drugs." Accordingly, people sickened by poultry products "might need to try several antibiotics before finding one that works," said *Consumer Reports* ("Dirty Chickens").

A report published by Food & Water Watch in 2007 said that 70 percent of all antimicrobials used in the United States are fed to livestock, accounting for "25 million pounds of antibiotics annually, more than eight times the amount used to treat disease in humans" (Muirhead 2007). Too many antibiotics can weaken the human immune system, as well as the immune systems of birds and other animals, increasing one's susceptibility to food poisoning and other illnesses. Moreover, food-borne bacteria such as *Salmonella* and *Campylobacter* don't necessarily just "go away." They can migrate from people's intestines to other body parts—blood, bones, nerves, organs, and joints—to cause seemingly unrelated diseases that may emerge only later in life, such as arthritis (Stanley).

Plans are not underway to reduce the forced rapid growth, crowding, and stress responsible for making birds sick. Chicken houses are larger and more densely crowded than ever, and they cannot be made clean. Every part the house, as well as the bird's own body, is a haven and breeding ground for disease organisms.

Now, as in the 1990s, only superficial solutions are promoted—food irradiation, trisodium phosphate carcass rinse, chlorine—"by far the most commonly used carcass and equipment disinfectant in the poultry industry in the USA" (Russell and Keener)—and other fake fixes summarized in the U.S. Department of Agriculture's "Hazard Analysis at Critical Control Points" plan known as HACCP, ordered by President Bill Clinton in 1996 (Weiss 1996). Real solutions are rejected as uneconomical. As a USDA official told a poultry symposium in 1987: "We know more about controlling *Salmonella* than we are willing to implement because of the cost factor" (Dubbert).

Once poultry products leave the plant, it is up to the retailer, food handler, and consumer to deal with the contaminated carcasses by following strict government guidelines that instruct people "to behave as if they're decontaminating Three Mile Island" just to have a meal.

Nevertheless, it is possible that in time some of these food safety problems will be brought under some sort of control, or seeming control, sufficient to satisfy public concern. Food irradiation will go into effect (Smith 2007), and "public education" will be hailed, though nothing will have changed in essence. Meat eaters will continue to eat flesh from infected sources only to be "cleaned up" at the end. However, society will feel that the contamination problem, and maybe even the animal welfare problem, has been solved. If this happens, why, then, should anyone care what happens to a chicken?

Chapter 5

The Death

The last truck pulled out about 3 a.m., the crates packed with quiet, huddling birds. The barn was silent, empty, strewn with smothered chickens, empty cans and bottles, occasional piles of human excrement mixed with the birds.' I was only thankful that we didn't have to witness the slaughter.

—Cathryn Baskin, "Confessions of a Chicken Farmer"

Do you think from your perception that the birds have a sense of what is going to happen to them?

Yes. They try everything in their power to get away from the [killing] machine and to get away from you. . . . They have been stunned, and so their muscles don't work, but their eyes do, and you can tell by them looking at you that they're scared to death.

—Virgil Butler, former Tyson chicken slaughterhouse worker, press conference, 2003b

The feet go to China, the intestines to the Turkish community in Germany, the legs to Eastern Europe, the wings to Spain, and the bones to soup-makers in Holland. The rest is packaged for the Irish and British markets.

—Ronan McGreevy, "The True Cost of Cheap Chicken," *Irish Times,* 2008

The death toll of chickens bred for the table exceeds all of our other killings of warm-blooded animals conducted for this purpose. Of the 8 billion animals slaughtered in 1995 in federally inspected facilities in which 99 percent of all birds killed for food in the U.S. are slaughtered, 7.8 billion were birds. Of these, 7.5 billion were chickens (USDA-NASS 1996a, b).

Of the 10.5 billion animals slaughtered in 2006 in federally inspected facilities, 9,252,320,000 were chickens, turkeys, and ducks. Of these, 8,968,916,000 were chickens, including 8,837,755,000 chickens raised for meat. The remaining 131,161 million chickens were "spent" fowl, including hens used to produce eggs for human consumption, and roosters and hens used for breeding.

The number of ducks slaughtered in USDA facilities in 2006 was 28,081 million. The number of turkeys, including both young birds and parent flocks, was 255,323 million (USDA-NASS 2007a).

To get an idea of the number of chickens being slaughtered compared to other animals in the United States, 33.7 million cattle were slaughtered in 2006 in federally inspected facilities, whereas approximately 35 million chickens are slaughtered *every day* in this country (USDA-NASS 2007b). Europe slaughters over 4.4 billion broiler chickens each year, Canada slaughters 900 million, and in 2006, Great Britain slaughtered 860 million chickens (Chalmers). Worldwide, 52 billion birds were reported slaughtered in 2005, 48 billion of them chickens, including egg-laying hens (*WATT Executive Guide 2006/07*, 10-14).

When we add to the 9 billion chickens slaughtered in the U.S. in a single year the 250 million or more male and defective female chicks destroyed at the hatchery by the U.S. egg industry and the 42.5 million chickens condemned postmortem at the slaughterhouse for airsacculitis (a ubiquitous *E. coli* respiratory infection in factory-farmed birds), septicemia (blood poisoning, the largest cause of slaughterhouse condemnations), tumors, being scalded alive, and other causes, we begin to form an idea of the death toll of chickens for human consumption (USDA-NASS 2007a).

Manual Chicken Catching

Typically, crews of seven to ten men catch and crate 7,000 to 10,000 birds per hour at a cost of approximately $20.00 per 1,000 birds per hour. As the men tire during eight-hour shifts, they slam birds against the cages, breaking wings and legs. A catcher may lift five to ten metric tons of chickens during an average night.

—Pamela Bowers, 1997, and Scott Kilman, 2003

At the chicken slaughterhouse, each day thousands of birds are crammed inside crates stacked on trucks, waiting to be killed. Truckload after truckload pulls into the loading dock. During the summer, huge fans rotate to reduce the number of birds who will die of heat suffocation while waiting to enter the plant. During the winter, an untold number of birds freeze to death in the trucks. Others fall out and freeze to the ground on the docks or somewhere along the way. A forklift picks the topmost pallet of crates off each flatbed truck, and the birds disappear into the darkness.

They came out of the darkness. "Live haul involves hand catching the birds, mostly at night, in a darkened, dust-laden atmosphere," a U.S. Department of Agriculture manual explains (Brant). Every night in the United States, approximately 8,000 chicken catchers put on throwaway suits, rubber gloves, and dust masks. In a few hours, the masks will be soaking wet and black with dust, and the men will tear them off in order to breathe. The lights of the 600-foot-long barns are extinguished. "And in a minute, the tranquil scene—quiet barns, frogs peeping in a pond to the night sky—is shaken into screeching, roaring chaos," as the men move into the barn clapping their hands and shouting to make room for a forklift with a five-foot cabinet of cages (Clark 1992).

When the forklift drops a box of plastic cages, the men crouch to a group of cornered chickens, groping for their legs, trying to grab in each hand one leg each of four or five birds who desperately fight back, wildly flapping their wings and pecking. For this they get their heads bashed against the ground. When a catcher thinks he has the right number of chickens in each hand, he pulls open a drawer in the cage and flings the birds in, pushing them down and shoving them into the back of the drawer. He stuffs in protruding body parts—wings, legs, heads—and slams the drawer shut. In three minutes, the drawers are full, and a forklift brings another empty cabinet. The work continues through the early morning light, until the house that held thirty thousand terrorized baby birds, so young they would still have been sleeping warm under their mother's wings, is empty and silent.

A postmortem examination in Britain of 1,324 broiler chicken carcasses of birds who arrived dead at the slaughterhouse found crushed skulls, presumed to have occurred when the drawers were closed. The investigators found abdominal hemorrhage and dislocated and broken hips, indicating that "catching and carrying large birds by one leg is conducive to dislocation of the hip, and that catching and carrying by two legs would help reduce this problem. Catching by two legs, however, would slow down the catching rate" (Gregory and Austin).

Theoretically, catchers are supposed to strive to reduce the bruises, broken bones, and smothering that result in lost profits; however, chicken "stuffers" are paid for speed, not gentleness. A former chicken "grower" writes, "By the dust-dimmed light of their flashlights, we watched as the crew hollered and yelled, trampling the frenzied birds indiscriminately.

I will admit that I didn't always treat the birds with the greatest of gentleness, but I was sickened as birds were kicked and thrown across the darkened barn. We had to leave" (Baskin). A reporter who stood shift with a crew remarked that the climate is not only frenzied and filthy, but angry. A crew leader states, "All of this job is bad. . . . You hate the work" (Clark 1992).

"Spent" laying hens are simply torn from the battery cages to the transport cages without regard to preserving "carcass quality." A former egg-industry worker in Britain explained, "I recall being shouted at for my gentleness. Birds were dragged from the cages by their legs. Four birds were carried in each hand end down, down the shed to the door. The noise was deafening, the smell was putrid. Legs, wings, and necks were snapped without concern. . . . I gave up work in the poultry industry after bad dreams at night" (Druce 1989, 1).

Automated "Harvesting"

Out of the gloom and dust of a chicken house as long as a football field, a Ph2000 emerged. . . . The nine-ton, 42-foot-long contraption crept closer, slowly sweeping a low metal ramp back and forth through the flock like a giant scythe.

—Scott Kilman, *Wall Street Journal*, 2003

Since the 1960s, in an effort to reduce the bruises, broken bones, and other injuries that result in market downgrading, and to get rid of the labor problem, the broiler chicken industry has sought to replace manual catching with automated catching machines. The current model looks like a "combination airport baggage carousel and tank" (Kilman).

Proponents claim that machine catching could reduce some of the injuries incurred by the approximately 25 percent of birds hurt during manual catching—a low figure one may assume, given that the birds arrive at the slaughter plants with ruptured livers, crushed heads, dislocated necks, hemorrhages from fractured femurs, and other wounds. These fragile, "overgrown baby birds, easily hurt," are "treated like bowling balls," as one poultry scientist put it (Webster 2002).

Supporters say that catching machines move "gently" through the chickens, scooping up 150 birds a minute, 7,000 to 8,000 birds an hour. Rubber fingerlike projections "nudge" the birds in their chests. When they lift their feet to get out of the advancing machine's way, the birds are forced to hop onto the ramp. As more birds are forced onto the ramp, they crowd one another toward a conveyor belt that blasts them with compressed air up the belt into a metal chute and into a crate (Kilman). When a crate is full, a forklift loads it onto the transport truck.

This machine replaces a pneumatic, "vacuum cleaner" model of the late 1970s, in which birds were handfed into a funnel-like aperture and sucked through tubes to crates on a truck. The birds clogged inside the tubes and were spewed out. Michael Lacy, a poultry researcher at the University of Georgia, found this amusing: "a neat fountain of chickens shootin' up into the air."

Reports vary on automatic catching machines. In one study, injuries, especially of the legs, were said to be fewer after mechanical catching, but the number of birds arriving dead at the slaughter plant was the same as with manual catching (Knierim and Gocke). Other studies have found that bruising, fractures, and preslaughter deaths were increased by machine catching (Turner, 22).

Former Tyson employee Virgil Butler said birds caught by machine where he worked arrived at the slaughter plant "beat up, bruised up, had broken bones, and were generally broken apart." The amount of filth on the machine-caught birds was greater than with hand catching, because the machine "swept up hundreds of pounds of dust, dirt, chicken litter, lime, etc." along with the birds, whose rate of heart failure at the plant increased with machine catching, he said (2003d). One farmer wrote:

> Chickens aren't going to line up in front of the counter rotating rotors with soft, fingerlike projections to be jostled into crates for The Final Journey. Instead, they're going to be spread out all over the broiler house, making a fuss. You would too if, suddenly, one end of your house opened up and there was this monster machine, 12 1/2 feet wide, 24 feet long, and 8 feet high, with three giant rotating "claws" coming at you at speeds "up to 18 miles an hour" (Hotchkiss).

An organic chicken farmer in British Columbia said the machine "is cruel as far as I'm concerned. They've had devices that used to vacuum birds up, shooting them out the back and breaking their wings and legs. It was a nightmare. You can't make a machine that won't hurt them" (Bellett).

In 2003, 5 percent of chickens were claimed to be caught by machine in the U.S., according to the *Wall Street Journal* (Kilman). Machine catching had not yet come to Canada, and manual catching continued to be the main method used in the European Union (Turner).

Transportation

It was found that half the birds that arrived dead at the plants had died from heart failure. . . . Presumably the physiological responses associated with the stress of catching, loading, and transporting the birds have been too much for the cardiovascular system to cope with.

—Gregory and Austin, 502

Broilers that are dead-on-arrival at processing plants are a common occurrence throughout the poultry industry.

—"Minimizing Mortality," 2004

Ten to twelve chickens weighing four to five pounds or more each are forced to occupy three and a half square feet of cage space in the transport trucks. Many birds die in the trucks awaiting the traumatic trip to the slaughterhouse following catching. Others die en route and on the loading docks at the slaughter plants. To reduce the number of birds dying in the summer heat, they are manually hosed down while the trucks are parked near large fans. Poultry transport, which can last anywhere from half an hour to 12 hours or longer, takes place in all kinds of weather, typically in uncovered trucks. On the Eastern Shore of Virginia where I live, some trucks taking chickens to slaughter are boarded up during the winter.

Clare Druce of Farm Animal Welfare Network said, "Half-naked battery hens will feel cold winds, especially." According to *Diseases of Poultry*, "Unless crates are properly covered, exposure to wind and cold will rapidly cause freezing of unfeathered parts. The frosted appendage first becomes red and swollen, followed by gangrene, necrosis, and sloughing. After the appendage thaws, the bird experiences intense pain" (Hofstad, 744).

In January 2007, Animals' Angels, a group specializing in farmed animal transport investigations, documented turkeys being trucked to slaughter in Iowa's freezing winter:

> We picked up and trailed a densely loaded turkey truck just outside of Des Moines, Iowa. We followed the truck for four hours to its destination—Sara Lee Foods in Storm Lake, Iowa. The truck was without tarps (as were all Sara Lee trucks carrying turkeys), affording the birds no protection from the cold and wind.

> There was a large amount of blood on the ceiling of one of the cages. Once the trailer arrived at the plant (at 6:30 p.m.), the turkeys still had another two-hour wait in the open cold as the truck had to queue up to enter the facility. . . . When we expressed our concern about the birds being exposed to the cold and developing frostbite on their legs and feet, they responded: "We cut those parts off."

Stress and death among birds in transport are so common that the causes are studied in laboratory simulations. Spent laying hens have been shown "to experience a level of fear comparable to that induced by exposure to a high-intensity electric shock" (Mills and Nicol). Broiler chickens are so averse to the truck vibrations that the stress and muscle tension they experience impairs their ability to regulate their body temperature, increasing the severity of heat stress (Turner, 23). U.S. researchers call thermal stress and physical injury "leading causes" of birds arriving dead at the slaughter plants ("Minimizing Mortality").

Heat stress is a major problem in poultry transport and holding. Temperatures at 48 to 50 degrees F (9 to 10 degrees C) when the birds are loaded in the transport crates will climb to 55 to 60 degrees F (12.5 to 15.5 degrees C) in transit and up to 85 degrees F (30 degrees C) when a loaded truck stops for as little as an hour before unloading the birds at the plant (Phelps).

Imagine the suffering of chickens being trucked from, say, an egg farm in Arizona through the Arizona desert, where temperatures, even at midnight in the summer, are over 100 degrees F (38 degrees C) outside a vehicle, and double that inside, to slaughter in California, at least an eight- to nine-hour trip.

It is claimed that some heat stress problems could be solved by attaching an air scoop to vehicles; however, this solution has been rejected because scoops would increase wind resistance and drive up fuel costs. At the Roslin Institute in the U.K., a controlled-ventilation chicken transport truck launched in 2000 was said to be "the most advanced design for the movement of live birds" (AATA). However, this vehicle does not appear to be widely used.

Even if some solutions to heat stress and cold are eventually implemented, the birds' need for food, water, and rest cannot be met. Rest would merely prolong the journey, and there is no feasible way to provide food and water for thousands of birds crated and stacked in a truck. The traumatized birds couldn't eat or drink anyway under these conditions. Moreover, food and water are deliberately withdrawn from broiler chickens and turkeys from one to four hours or more before catching, in order to reduce intestinal splatter at the slaughterhouse, and from battery-caged hens several days before catching to save money on hens deemed totally worthless (Webster 1996).

Removal of food and water from the birds prior to transport disrupts their gastrointestinal tract and impairs their immunity, increasing the number of birds infected with *Salmonella* 10 times above the number of birds infected before catching (Fliss). Food deprivation increases *Salmonella* and *Campylobacter* contamination, not only because of the stress on the birds' immune systems, but because they are driven to peck at the contaminated, manure-soaked litter they're bedded in after their food has been taken away ("Pre-harvest Interventions").

Truck Accidents

Trucks carrying thousands of chickens are liable to overturn on the highway. For example, on a freezing morning in January 1993, 5,000 chickens on the way to slaughter outside Portland, Oregon, fell 50 feet from a highway ramp onto a parking lot when the truck carrying them hit a guard rail. Instantly, 2,000 birds were killed. The rest were gathered up and sent on to the slaughterhouse. This event was treated as a joke by the local newspaper and the company (Painter).

The accident that I became involved with occurred on August 24, 1995, when an 18-wheeler carrying 5,000 spent broiler breeder hens and roosters, traveling from North Carolina to a slaughter plant in New Jersey, overturned on Interstate 95 in Springfield, Virginia, crushing 1,000 birds instantly (Shear). Hundreds of crying, terrified birds covered the highway as crews roughly grabbed them by their wings and kicked them in front of TV cameras. Others sat locked in the jumbled crates in 90-degree F (32-degree C) heat, unable to move. Police said it was against the law to rescue the "merchandise." As a result, about half the birds died of heat suffocation in the crates next to the highway. Hundreds of others were given lethal injections during the night. Except for the 16 hens we sneaked into the car, the rest were loaded back onto a truck and sent on to slaughter (Fountain).

No Federal Regulations

There are no federal laws in the United States regulating poultry transport. The Animal Welfare Act excludes transportation of animals used for food and fiber, and the Twenty-Eight Hour Law of 1877, requiring animals in interstate transport to be given food, water, and rest every 28 hours, has historically been applied only to trains and ships, not trucks or planes, and it excludes farmed animals transported within a state, thereby excluding millions of chickens raised in complexes in the same state as the slaughter plants they're going to.

The Twenty-Eight Hour Law carries a maximum penalty of $500 and isn't enforced anyway. In 2006, following a legal petition filed by several animal protection groups (HSUS 2006), the U.S. Department of Agriculture reversed its policy of excluding farmed animals from protection during long-distance truck transport—yet, egregiously, the USDA continues to ignore the nearly 10 billion chickens, turkeys, and other birds being trucked to slaughter each year in the United States.

Great Britain has a 15-hour-transport limit, and Europe has an 8-hour limit for standard vehicles, with exceptions similar to those in the U.S. (Wolfson). In Britain, the Welfare of Animals During Transport Order of 1992 theoretically could enable local authorities to prosecute poultry transport companies for keeping birds on lorries (trucks) for more than 12 hours at a time. Officers could point to a requirement in the legislation that birds must be given water every 12 hours, but this is impractical. According to Farm Animal Welfare Network, the Order, which was designed to placate European Convention lawmakers and U.K. animal welfarists, is worded so confusingly as to make it unlikely to bring about any improvement for birds ("Poultry Transport—Unimproved?").

Shipment of Baby Chicks

The real romance of modern poultry husbandry has been the unprecedented growth in the production and shipment of readymade baby chicks. Hatched in mammoth incubators on breeding farms or at commercial hatcheries, the chicks provide the most economical and convenient method of securing one's foundation stock, of enlarging one's flock, and of providing future generations of layers. . . . [S]ome 25 years ago, from the little village of Stockton, New Jersey, in the Delaware River Valley, the first baby chicks were shipped.

—Harry R. Lewis, "America's Debt to the Hen," *National Geographic*, April 1927

Live day-old poultry—chickens, ducks, geese, partridges, pheasants, guinea fowl, quails, and turkeys—are shipped through the mail, including airmail. They are handled exactly like luggage—the cheapest way, as established by post office rules in 1924. U.S. postal regulations require only that the birds—the "perishable matter"—be delivered to the receiver within 72 hours of hatching, with no provisions being made for food, water, or weather. As a result, millions of baby birds are delivered dying or dead each year. Unclaimed birds may be left to die, suffocated in plastic bags, and otherwise cruelly disposed of. Postal workers who find boxes of dead and dying baby birds shipped through their facility are forbidden to interfere. Hatcheries ship baby roosters to customers as packing material, referring to these throwaway birds as "packers."

In 2001, Northwest Airlines joined United Airlines and American Airlines in refusing to carry baby birds as mail, after 300 chicks in boxes bound for Ohio were left out in the rain in Minneapolis. At the time, Northwest told the *Wall Street Journal* that between 60 and 80 percent of baby birds die on some flights, "often because of excessive heat or poor packaging by hatcheries." As well, many birds get crushed during flights, and they often freeze to death or sit unattended in mailrooms on Sundays and during holidays (Spurgeon and Power).

Commercial hatcheries claim that newborn chicks can go without food or water for 72 hours (three full days) after hatching, because, in nature, when chicks hatch with a mother hen, the earliest hatched chicks must wait for all of the chicks to hatch. They survive by absorbing their yolk nutrients during this time.

However, in nature, a clutch of chicks normally hatches within 24 hours, or at most within 48 hours—not 72 hours. Chicks kept for up to 48 hours are susceptible to dehydration, according to the *Veterinary Record*, which observes that "in North America earlier hatching chicks could be held in the incubator for up to 36 hours after hatching." This is before the flight or ground transportation, which may include stop-offs, layovers, and other delays, has even begun (Warriss, 49).

According to a poultry industry manual, chicks shipped for breeding purposes over long distances typically lose "20 percent of their bodyweight before arrival at the farm" (Bell and Weaver, 630). Many chicks who survive the flight die upon reaching their final destination, because the stress and lack of food and water have left them too weak to eat and drink. Birds with dehydrated internal organs do not recover.

Thousands of Baby Turkeys Perish in Airline Transport

An example of what can happen to baby birds in airline transport occurred in July 2006. Some 11,840 turkey poults died while being transported by Northwest Airlines in two separate incidents that month. The first occurred on July 13th, when more than 9,000 of 11,500 poults, crowded onto a single flight from Detroit to San Francisco, died from suffocation, overheating, and dehydration. The birds were being transported from a Hybrid Turkeys farm in Canada to Zacky Farms in Fresno, California, for breeding purposes.

The second incident occurred on July 19, when some 2,240 baby turkeys sat for hours in 108-degree F (42-degree C) heat in Las Vegas after the Air Canada flight they were on developed mechanical problems. According to the Peninsula Humane Society, Northwest, which handles Air Canada's baggage, threw most of the birds, dead and alive, into a trash compactor (Nguyen).

Mass Transport Incompatible with Poultry Welfare

The mass transportation of birds is inherently cruel. As Compassion in World Farming has explained, as long as consumers demand the mass killing of chickens and other birds for food, the birds will "be manhandled, injured, covered in filth, hungry, and thirsty, and just plain terrified from the moment they are caught to the time of their death at the slaughterhouse" ("Sad Journeys").

The Slaughter: Dumping, Shackling, "Stunning," Throat-cutting, Bleeding, Scalding

Every chicken is bled out while still sentient. They hang there and look at you while they are bleeding. You can definitely tell that they know what is going on. Sometimes if they are not completely immobilized by the stunner (which happens frequently), they will try to hide their head from you by sticking it under the wing of the chicken next to them.

—Virgil Butler, "Clarification on Stunner Usage," 2004

On October 18, 1991, I toured a former Holly Farms (now Tyson) plant, which, at the time, slaughtered 200,000 broiler chickens a day. We began in a packing room filled with boxes of frozen chicken parts labeled for shipping to Russia and proceeded through the various stages to the loading dock where birds who fell onto the cement were pulled back by workers using long-handled hooks. We stood beside the shackled baby birds in the dark red room designed to "calm" them in preparation for the intense suffering that awaited them in the electrified water-bath "stunner," and beyond.

—Karen Davis 1996, 166

In the summer of 1990, I stood outside the Perdue Farms chicken slaughterhouse in Salisbury, Maryland, and watched the trucks, each one stacked with thousands of chickens, roll in and out all day. Across the street, people drifted in and out of McDonald's, some no doubt to dine, others just having dined, on Chicken McNuggets.

At the slaughterhouse, birds may wait in the trucks from one to nine hours or longer, depending on the killing and processing speed. It is a throat-catching moment to look at a truck stacked with orange plastic crates that seem empty, and all of a sudden see movement of an eye in there and know that experiences are taking place inside.

Live-Hanging Cage

I asked Virgil to take me back to the hanging cage where he worked so I could see for myself what it looked like. I had prepared myself to feel disgusted, sad, and uncomfortable, but nothing could have prepared me for the way I felt when I saw it. The only thing I can even try to compare it to would be that feeling you get in places like hospitals and jails, where there is suffering and death, dread and fear. Take that feeling and magnify it by at least 10 and you will have maybe an inkling of what I felt at the door of that room that day. I couldn't leave fast enough.

—Laura Alexander, "Slaughterhouse Worker Turned Activist"

Standing next to a truckload of chickens at a Tyson plant in Richmond, Virginia, I saw how agitated the birds became as they watched their companions being yanked by the legs and shackled by their ankles upside down on the conveyer belt. An employee in Heavener, Oklahoma, wrote of his first day on the job in the live-hang area: "It was dark inside, with an eerie blue glow meant to keep the birds quiet. The hens would be dumped from the chicken truck cages outside onto the conveyer belt. There were so many chickens coming down the conveyer belt that there was not enough room for them all. Some would fall three feet down to the concrete floor. Within minutes of starting, my mask was full of fecal material and feathers. I was literally breathing in chicken excrement and feathers with every breath. The aroma was indescribable" (Stevens).

The crated birds are dumped onto a conveyer prior to stunning, a hard, painful banging that breaks so many wings that some industry scientists insist that dumping needs either to be phased out or else adapted to a gas-stunning system in which the birds would be moved through the stunner on a conveyer and dumped after stunning, then hung from the shackles in a more or less unconscious condition (Grandin 2005a). According to Mohan Raj, the force of the hanging procedure causes pain "because the affected bone surface is enriched with pain receptors causing over 90 percent of birds to flap their wings due to pain. There is enough scientific evidence to suggest that shackling live birds increases the prevalence of dislocated joints, broken bones, and muscle bruising in conscious birds" (Raj 2004).

"One-Leggers"

Along with birds hung "correctly," with both legs clamped in the shackles, Temple Grandin has reported seeing "a lot of one-legged shackling" (Grandin 2002). "One-legger" is the term used to describe a chicken who has accidentally been hung in a shackle by only one leg. Former Tyson employee Virgil Butler described how such chickens at the plant where he worked in Grannis, Arkansas—between 200 and 1,500 birds per nightshift—were ripped out of the shackle, tossed in a pile, and left to die, since it is the live hanger's job to try to keep one-leggers from entering the kill room where, in their desperate struggle to free themselves with the one dangling leg trying to clutch at everything from workers to machinery to adjacent birds, these chickens can miss the stunner and get chopped up by the automated killing blade. Some birds, he wrote, will try to "flop back out in the floor and get kicked back into the corner. I have personally seen this happen six or more times to the same chicken before it finally bled to death" (Butler 2003c).

"Stunning"

Stunning is a procedure that induces an unequivocal pathological brain state that is incompatible with the persistence of consciousness and sensibility in order to perform slaughter without causing avoidable fear, anxiety, pain, suffering, and distress. . . . The stunning method itself should not be stressful.

—Mohan Raj, USDA Seminar, December 16, 2004

Birds slaughtered in the United States are neither stunned (rendered unconscious) nor anesthetized (rendered pain free). Preslaughter stunning is not required by law and is not practiced, despite the use of the term "stun" to denote what is really immobilization of conscious birds. In practice, "stunning" is monitored only for efficient bleedout. The Poultry

Products Inspection Regulations of the U.S. Department of Agriculture merely states, "Poultry shall be slaughtered in accordance with good commercial practices in a manner that will result in thorough bleeding of the carcasses and assure that breathing has stopped prior to scalding" (PPIA).

There are three main methods for immobilizing birds to prepare them for slaughter, or neck-cutting: (1) chemical immobilization, in which a mixture of gases is administered, such as carbon dioxide and reduced oxygen, using an inert gas such as argon or nitrogen to stabilize and improve dispersal of the main gas; (2) mechanical, as by debraining, in which the medulla of the brain is pierced directly through the eye, a traditional farming practice prior to the development of electrical stunning in the 1930s; and (3) electrical, the standard commercial method in which a live current is driven through the bird by means of an electrified knife or plate, or electrified water to which sodium chloride (salt) has been added to improve conductivity of the charge.

Preslaughter Electrical Water-Bath "Stunning"

The complexity of multiple bird water-bath stunning is not conducive to maintaining good welfare. Effectiveness of the stun cannot be determined. The method, widely practiced because it is simple and cheap, cannot be controlled. You can't control the amount of electrical current flowing through a bird. You can't harmonize electrical resistance in broiler chickens. The water bath has to be replaced.

—Mohan Raj, USDA Seminar, December 16, 2004

The electrified, cold-salted water bath is the standard method used in the large commercial slaughter plants to immobilize birds prior to cutting their throats. The method was developed in the twentieth century to perform strictly commercial functions rooted in farming practices such as those described in a 1937 manual, *Marketing Poultry Products*, by Benjamin and Pierce, who wrote: "It is necessary that the brain be pierced with a knife so that the muscles of the feather follicles are paralyzed, allowing the feathers to come out easily" (139).

After the birds have been manually jammed into the moveable metal rack that clamps them upside down by their feet in the "live-hang" room, they are dragged through a 12-foot-long electrically charged, water-filled trough, called a stun cabinet, for approximately seven seconds. Between 20 and 24 birds occupy this cabinet at a time. Every minute, 180 or more birds pass through it (Bilgili 1992, 139).

The electrically charged water bath is not designed to render birds unconscious, or even pain free, but to slacken their neck muscles and contract the wing muscles for proper positioning of their heads for the automatic neck-cutting blades. It is also designed to prevent excessive struggling of the birds as the blood drains from their necks, to promote rapid bleeding (in under 90 seconds), and loosen the birds' feathers after they are dead. During electrical water-bath stunning, currents shoot through the birds' skin, skeletal breast muscles, cardiac muscles, and leg muscles, causing spasms and tremors, reducing their heartbeat and breathing, and increasing blood pressure. The birds exit the stunner with arched necks, open, fixed eyes, tucked wings, extended rigid legs, shuddering, turned up tail feathers, and varying amounts of defecation (Bilgili, 136, 142).

Problems identified with this method include birds missing the stun bath by raising their heads to avoid it (as, for example, in the case of the "one-leggers" described above), and shocking of birds splashed by water overflowing at the entrance end of the stun cabinet. Electrical resistance of the circuits can vary between and within a single slaughter plant, reflecting differences in stunners and circuits and a wide range of other variables including the birds' own bodies. According to Bilgili, "The abdominal fat tissue has the greatest resistivity of all tissues measured. The high variation observed in resistivity of the skull bone indicates that birds with thick and dense skull bones [spent laying hens and breeding fowl, because of their age] are most likely to be inadequately stunned" (140–141).

Evidence notwithstanding, the U.S. Department of Agriculture's Food Safety and Inspection Service (FSIS), in charge of the approximately 320 poultry slaughter plants under federal inspection in 2006, claims that most birds under inspection are slaughtered "humanely." However, FSIS does not keep a list of "humane" methods or provide documentation verifying that most birds are rapidly and effectively rendered insensible to pain and suffering in the process of being killed. H. Russell Cross, FSIS Administrator in 1992, wrote in response to my inquiry: "The statement that 'birds are effectively stunned before slaughter' is based on observations of Food Safety and Inspection (FSIS) personnel."

An example of a published FSIS study is "A Survey of Stunning Methods Currently Used During Slaughter of Poultry in Commercial Poultry Plants" (Heath 1994). Cited at a Congressional Subcommittee hearing in 1994 as showing "widespread use of humane methods of slaughter in the Nation's [poultry] slaughter plants," this 1992 survey was conducted entirely by phone and fax!

Birds are Not Stunned

The typical amperage used in stunning by our pulsating direct current pre-stunner is approximately 12 to 15mA . . . . If the reading is 200mA, with 16 birds in contact, there would be an average of 12.5mA per bird.

—Wayne Austin, Simmons Engineering Company

In reality, so-called humane electrical stunning of poultry is regarded as incompatible with the goals of commerce. High levels of current are said to interfere with plant efficiency and to cause hemorrhage—a "bloody bird" (Kuenzel). Hemorrhaging of the fragile capillaries of the increasingly younger and heavier birds being slaughtered has been cited as a reason to lower the electrical currents even more (Bowers 1993b). While research suggests that for electrical stunning to produce unconsciousness chickens should receive a minimum current of 120 mA per bird, and that currents under 75 mA per bird should never be used (Gregory and Wotton, 219), chickens slaughtered in the United States are being given weak currents ranging between 12 mA and 50 mA per individual bird. As researcher Bruce Webster told a conference on handling and stunning, "Industry is trying to stay at 25 mA and below due to hemorrhaging" (Webster 2002).

Electrical Paralysis of Conscious Birds

For death to be painless and distress free, loss of consciousness should precede loss of motor activity (muscle movement). Loss of motor activity, however, cannot be equated with loss of consciousness and absence of distress. Thus agents that induce muscle paralysis without loss of consciousness are not acceptable as sole agents for euthanasia.

—AVMA, 2000 Report of the AVMA
Panel on Euthanasia, 675

Since physical signs like the absence of breathing (apnea) are the same in properly and improperly stunned birds, these signs cannot accurately indicate the subjective condition of an electrically "stunned" bird.

—Mohan Raj, USDA Seminar, December 16, 2004

All it [the stunner] does is paralyze the muscles. . . . In Tyson's own words to the workers, "It makes the plant more efficient." They never said anything about "humane."

—Virgil Butler, "Clarification on Stunner Usage," 2004

According to researchers, a major problem with electrical stunning, even under "ideal" conditions, is that birds who are stunned (rendered unconscious) and birds who are merely paralyzed look the same (Gregory 1986; Boyd, 224). A bird or a mammal may be unable to move, struggle, or cry out while experiencing intense pain and other forms of suffering including the inability to express outwardly a response to pain perception. As Nobel laureate A. V. Hill explained about the electrocution cabinets being used for dogs and cats in the 1920s, these cabinets "were likely to cause great pain, although this would be masked by muscular paralysis." The apparently unconscious or dead animal was most likely to be "fully conscious and in agony for some time before unconsciousness and death supervened" (WFPA, 12).

This gives an idea of what is happening to billions of birds in the slaughterhouses—bearing in mind that unlike the dogs and cats cited above, poultry are not intended to be electrocuted (killed) but only shocked into paralysis. Even after a century of controlled laboratory experiments, scientists disagree over how to determine whether a bird is truly stunned and not merely immobilized (paralyzed) and whether a bird is in pain. Various indicators have their proponents: visual, auditory, evoked versus spontaneous somatosensory response, physical activity, brain waves, breathing, and so on (Boyd). Imagine the feelings of the chicken or turkey for whom the following is recommended: "A good rule of thumb for checking for an adequate [electrical] stun is to remove the bird immediately after the stun and place it on the floor. The bird should be able to stand within one to two minutes" (Wabek).

British law requires that livestock and poultry must be rendered instantaneously insensible to pain until death supervenes. Poultry slaughter expert Neville Gregory said the law should delete the reference to pain and simply read, "rendering the animal instantaneously insensible until death supervenes," because following electrical "stunning," in addition to the suffering induced by the electric shocks, one can have analgesia, whereby there is conscious perception of nonpainful but highly distressing stimuli including gagging, breathlessness, smell of blood, fear, and apprehension (Gregory 1993).

In other words, one can have dreadful experiences even without being in physical pain. An example is the experience of breathlessness, or dyspnea. According to the American Pain Society, "The word 'dyspnea' subsumes a variety of unpleasant respiratory sensations described by terms such as chest tightness, excessive breathing effort, shortness of breath, and air hunger." The Society further states the following:

> There are few, if any, more unpleasant and frightening experiences than feeling short of breath without any recourse. . . . For instance, a strong perception of a need to breathe causes diving animals to surface and causes all animals to struggle to remove external obstruction to air passages (Banzett and Moosavi, 1–2).

Responding to welfare concerns, some European companies have begun subjecting chickens and turkeys to amperages designed to provoke cardiac arrest—a heart attack—in order to induce brain death prior to neck cutting and bleedout. Stopping the heart interrupts the flow of oxygenated blood to the brain, resulting in a presumed loss of consciousness. Birds in a state of cardiac arrest may be further protected from the protracted agony of badly cut necks. Notwithstanding, as one slaughter operator states, "It is possible that the [electric] shock, even as it renders the bird unconscious, is an intensely painful experience" (Boyd, 223).

Postslaughter Electrical "Stunning"

In addition to preslaughter "stunning," postslaughter electrical shocking of the still-living birds is being experimentally and commercially conducted. U.S. researchers claim that while it will not improve bleedout, it will "calm [the bleeding and dying] birds and reduce the force required to remove feathers" (Bowers 1993a). According to an article in *Poultry Marketing and Technology*, "Postslaughter stunning is mostly used on broilers weighing more than 7 pounds, light and heavy fowl, and turkeys. It is also recommended for processors cooking product for frozen entrees" (Bowers 1993b).

Thus, hanging and dying in the bleedout tunnel, after having their throats cut, the battered birds are guided automatically against an electrified ladder or a square plate and delivered a few final volts of electricity.

Neck Cutting and Bleedout

Some nights I worked in the kill room. The killer slits the throats of the chickens that the killing machine misses. You stand there with a very sharp 6-inch knife and catch as many birds as you can because the ones you miss go straight into the scalder alive. You have to cut both carotid arteries and the jugular vein for the chicken to die and bleed out before hitting the scalder. . . . The blood can get deep enough to go over the top of a 9-inch set of rubber boots. I have seen blood clots so big that it took three big men to push them. You have to stomp them to break them up to get them to go down the drain. That can happen in just 2 1/2 hours. We filled up a diesel tanker truck with blood every night in one shift. I have actually had to wipe blood clots out of my eyes.

—Virgil Butler, "Slaughterhouse Worker Turned Activist"

The two methods most commonly used for cutting the blood vessels in the necks of chickens and turkeys are manual cutting, in which a knife is passed across the side of the neck at the joint with the bird's head, and automatic neck cutting, in which the bird's neck is glided across a revolving blade—"a 6-inch meat saw blade that resembles a finishing blade for a circular saw" (Butler 2003c). Plants with automatic neck cutters may or may not have a manual backup, should a bird miss the cutter. Britain passed a law in 1984 requiring manual backup of automatic cutters. However, there is no law in the United States.

The fastest way to produce brain death in chickens by neck cutting is severing the two carotid arteries that supply the brain with most of its fresh blood, whereas the jugular veins carry spent blood away from the brain. Poor neck cutting extends the time that it takes a bird to die. Worst is the severance of only one jugular vein, which can result in a bird's retaining consciousness, while in severe pain, for as long as eight minutes. Most of the blood has to drain out of the body before the heart stops pumping blood to the brain through the carotid arteries. If both jugular veins are cut, brain failure occurs in approximately six minutes, and the bird is in danger of regaining consciousness, especially if breathing resumes. If both carotid arteries are quickly and cleanly severed, the supply of blood to the brain is disrupted, resulting in brain failure in approximately four minutes. However, the carotid arteries are deeply embedded in the chicken's neck muscles, and even more deeply embedded in the turkey's, making them hard to reach (Gregory 1984).

Cutting the spinal cord is regarded as inhumane because it induces asphyxia—suffocation—rather than depriving the brain of blood, because the nerves that control breathing are severed within the spinal cord. Cutting the spinal cord interrupts the nerves connecting the brain with the bird's body, making it impossible for the bird to exhibit conscious awareness through physical expression. As with the use of electricity and paralytic drugs, a bird in excruciating pain or other distress will not be able to show it.

Slaughter without "Stunning": Ritual Slaughter, Live Bird Markets, and Small Farms

Small slaughterhouses and farms often forego "stunning" in order to save money and because ritual slaughter excludes the practice. Typically, the birds are killed "by cutting the neck and incising one or more major vessels" (Heath 1994). This is often done after the bird has been shoved upside down into a killing cone with the bird's head hanging out at the bottom. Instead of being electrically paralyzed, the bird is physically restrained. Jewish doctrine states than an animal must be uninjured at the time of killing, and stunning is classed as injury (Birchall, 46). The Vietnamese puncture a chicken's throat and let the blood drain out slowly (Huckshorn).

Undercover footage of an ethnic slaughterhouse in Los Angeles shows chickens having their throats cut manually and being stuffed alive into bleedout holes by the employees. Blood-soaked chickens with partially cut throats try vainly to lift themselves out of the troughs into which more bleeding and writhing birds are casually flung before being picked up and shackled. Bleeding, flapping chickens fall off the line onto the floor—no one pays any attention (Farm Sanctuary 1991).

The United Poultry Concerns video *Inside a Live Poultry Market* shows footage obtained at the Ely Live Poultry market in the Bronx in New York City. In the slaughter room, two pitiful brown hens stand together in a stainless steel sink while men slice the throats of other chickens and shove them into the bleedout holes. The dying birds' legs pedal and thrash violently in the air. One slaughtered hen leaps out of the hole, alive, onto the floor. After a while, one of the slaughterers picks her up and shoves her back down into a bleedout hole, like he was stuffing garbage into a trashcan (UPC 2003a).

Ritual Slaughter

According to my Koran, animals have no voice. But you treat them like you treat yourself.

—Riaz Uddin of the Madani Halal slaughterhouse
in Queens, New York (Drake)

Ritual slaughter refers to "a method of slaughter whereby the animal suffers loss of consciousness by anemia of the brain caused by the simultaneous and instantaneous severance of the carotid arteries with a sharp instrument and handling in connection with such slaughter" (*HMSA, Title 7 U.S. Code,* Section 1902b). Contrary to assertions, ritual slaughter (that is, kosher and Muslim) does not cause a humane death. Neck cutting, even if done "correctly," is painful and distressing (Raj 2004), and other problems have been identified. Researchers at the Food Research Institute in Britain showed that "in cattle, brain activity sometimes persisted for some time after shechita" (Jewish ritual slaughter), and that "sometimes the carotid arteries balloon within 10 seconds of being cut, causing an increase in blood flow to the brain, and so maintaining its activity" (Birchall, 46).

In practice, ritual slaughter—meaning the quick, clean severance of both carotid arteries carrying oxygenated blood to the brain—may not even be done. A New York State "Shopping Guide for the Kosher Consumer" states that the shocket [orthodox ritual slaughterer] severs the windpipe and jugular vein" (Ratzersdorfer 1987). At an Empire Kosher slaughter plant in Pennsylvania, owned by the largest kosher slaughter operation in the world, one worker removes the birds from the crate and passes them through the opening in the wall to another worker who holds and positions the bird for the slaughterer. The slaughterer severs the windpipe, or trachea, which is filled with pain receptors, and a jugular vein, and inspects the bird. The second helper then hands the bird to another worker who hangs it on a convener belt. A Baltimore, Maryland journalist who toured an Empire Kosher plant wrote that "the chickens thrash desperately on the hooks" but was told this was just "reflex" (Oppenheimer, 46). As for the footage obtained at the Ely Live Poultry market in the Bronx, Mohan Raj said it was "deeply disturbing to see the slaughterman restraining conscious birds by folding back their wings and cutting their throats as though he was slicing a fruit or vegetable" (Raj 2005a).

In addition to the cruel slaughter, the British Farm Animal Welfare Council reported the often "callous and careless" treatment of birds occurred in ritual slaughter markets, including throwing and ramming them into bleeding cones after their throats were cut and leaving rejected birds in transport crates overnight without food and water (Birchall, 46).

Spent Laying Hens and Small Game Birds

In the United Kingdom, spent laying hens are "stunned" using electrical water bath stunners or inert gas mixtures, and there is a quail plant in which the birds are hung on a purpose-built shackle line and dragged through an electrical water-bath stunner (Raj 2005b). In addition, a European Union Council Directive allows a vacuum chamber to be used "for the killing without bleeding of certain animals for consumption belonging to farmed game species (quail, partridge, and pheasant)." The birds are placed in an airtight chamber in which a vacuum is achieved by means of an electric pump. They are "held in groups in transport containers which can be placed in the vacuum chamber designed for that purpose" (Picket). Regarding the vacuum chamber, Mohan Raj said, "It can be extremely painful to birds with blocked ear canals" (2005b)

Research on vacuum stunning is being conducted on broiler chickens by the U.S. Department of Agriculture's Agricultural Research Service using a system claimed to be cheaper and faster than gas-based controlled atmosphere stunning (CAS) systems. According to an article, it takes "about 30 seconds to pull the pressure down to about 20 percent of normal atmosphere pressure in the chamber, or 80 percent vacuum. Within 30 seconds at 80 percent vacuum the birds are dead" (O'Keefe 2007, 25).

In the United States, spent laying hens may or may not be electrically "stunned" and small birds such as quails and guinea fowl normally are not. It is claimed that electrical stunning would incur a financial cost through carcass damage and rejection because of easily fractured bones (Boyce). According to Bruce Webster, spent laying hens "struggle in the shackle and lift their bodies away from the stunner bath, reducing the probability of making good electrical contact with the stunner." They flex their necks, get splashed with electrified water, struggle more violently with the additional pain, and ride up on the bodies of adjacent birds (Webster 2007). A USDA survey of slaughter-plant methods in 1991 showed that virtually all small birds, including spent laying hens, were "slaughtered without stunning by severing the carotid arteries or [by] decapitation. Light fowl (93 percent) and geese (100 percent) were slaughtered primarily by severing the carotid arteries. No geese were electrically stunned" (Heath, 299).

As spent laying hens are much older than broiler chickens when they are killed and consequently have harder skulls, they require stronger currents to make them unconscious (however that is determined), making the whole problem of electrical stunning even more insurmountable (Bilgili, 141).

GASSING

Gaseous stunning is intended to eliminate the problems inherent in multiple bird water-bath electrical stunning.

—Mohan Raj, USDA Seminar, 2004

The twin problems of bone breakage and pre-stun shocks make controlled atmosphere stunning (gas stunning) an option that fowl processors should seriously consider for spent commercial laying hens.

—Bruce Webster 2007

In view of the intense suffering caused by electrical stunning, an increasing number of researchers say that gas stunning based on hypoxia (low oxygen) or no oxygen (anoxia) represents the best alternative to electrical stunning of poultry. Gas would eliminate the need for preslaughter shackling. It could be performed in the transport crates, reducing stress on both workers and birds. Some say a nonaversive inert gas such as argon or nitrogen is the least inhumane method for spent laying hens (Raj 2005c). Others cite the difficulty of pulling hens in a state of rigor mortis out of battery cages. Yet another difficulty is evenly distributing the gas from floor to ceiling in a huge building full of stacked cages (Berg).

From a welfare standpoint, not all gases are the same. The two main gas systems recommended to replace electrical stunning are carbon dioxide (CO_2) and the inert gases argon and nitrogen. Carbon dioxide is the least expensive industrial gas, which unfortunately favors its use by the poultry industry. Speaking at a USDA Symposium in 2004, Mohan Raj explained the welfare advantage of argon and nitrogen to stun-kill the birds in the system known as controlled atmosphere stunning (CAS), or controlled atmosphere killing (CAK). The latter term was coined by People for the Ethical Treatment of Animals to stress the importance of killing the birds outright with the gas to prevent them from waking up during the slaughter process.

Carbon Dioxide

Birds and other animals completely avoid, hesitate to enter, or rapidly evacuate from an atmosphere containing high concentrations of carbon dioxide... Animal welfare scientists would agree that the humanitarian intentions of eliminating avoidable pain and suffering during water-bath stunning would be seriously compromised by carbon dioxide stunning of poultry.

—Mohan Raj," Recent Developments in Stunning and Slaughter of Poultry," 2006b

CO_2 activates brain regions in both birds and humans that are involved in the perception of pain. It causes panic in response to the sensation of suffocation and breathlessness, or dyspnea, which occurs when the amount of atmospheric CO_2 exceeds 30 percent. Inhalation of carbon dioxide is both painful and distressing because birds, like humans, have chemical receptors (intrapulmonary chemoreceptors) that are acutely sensitive to carbon dioxide. This sensitivity produces an effort to expel the gas by breathing more rapidly and deeply, but breathing more rapidly and deeply only increases the intake of CO_2, leading to suffocation. Ducks in particular have been shown to undergo an especially slow, agonizing death in CO_2 chambers (EFSA).

An article in *New Scientist* had a disturbing report by Ruth Harrison on the use of carbon dioxide. The author of the influential book *Animal Machines* and a member of the Farm Animal Welfare Council in Britain said, "I used to be very much a proponent of CO_2 stunning." But a visit to a mink farm in Denmark, followed by inhaling the gas herself, changed her mind. Regarding the gassing of day-old male chicks by the egg industry, which she once condoned, she said: "In my opinion, it is no better than the old practice of filling up a dustbin with them and letting them suffocate" (Birchall, 47).

Argon and Nitrogen

Both neck cutting without stunning and inhalation of carbon dioxide are "distressing and inevitably painful," according to Mohan Raj. In contrast, birds exposed to argon and/or nitrogen gases do not show aversion—they do not try to escape from or avoid the presence of argon or nitrogen, he says. The reason is that birds, like humans, have chemical receptors in their lungs that are acutely sensitive to CO_2, but they do not have receptors to detect argon, nitrogen, lack of oxygen, or reduced oxygen. Therefore, according to Raj, with these alternative gases they do not experience the pain, panic, and suffocation caused by exposure to CO_2.

In experiments in the United States and the United Kingdom, turkeys and chickens made fewer stops and retreats when argon was present, but they showed an increased tendency to stop when carbon dioxide was present. When CO_2 levels are high (above 40 percent), birds gasp, shake their heads, and stretch their necks to breathe. When they are in the presence of argon, birds exhibit no sign of gasping or stretching their necks to breathe. The small amount of head shaking in chickens in an argon chamber indicates that they are trying to "wake up," rather than experiencing suffocation, as in a CO_2 chamber, according to Raj.

However, birds do a lot of wing flapping that results in broken wing bones in the presence of argon or nitrogen. The poultry industry complains that broken wings can't be marketed to consumers (Grandin 2005b). From an ethical standpoint, wing flapping raises disturbing questions about the birds' suffering. However, Raj argues that wing flapping in the presence of these gases signals unconsciousness caused by the brain being starved of oxygen (Raj 2006b; Prescott 2006).

Adoption of Gas Systems

An amendment to U.K. regulations in 2001 allowed gas mixtures to be used in the slaughter of poultry. Approximately 25 European slaughter plants use some sort of gas system (Shane 2005). It is estimated that 75 percent of turkeys and 25 percent of chickens slaughtered for human consumption in the U.K. are killed using inert gas mixtures (nitrogen and/or argon) or a mixture of less than 30 percent carbon dioxide in the inert gases argon and/or nitrogen (Raj 2006b). Several major chicken and turkey slaughter operations, including Deans Foods, which slaughters 7,000 spent hens and breeding fowl per hour in Lincolnshire, England, use a nitrogen-based stun-kill system. Deans Foods calls it "the most welfare-friendly system of stunning poultry available" ("Stunning Advice").

There is more resistance to replacing electrical "stunning" with gas-induced stunning of birds in the U.S. than in the U.K. The U.S. industry views the issue as having to "counter opposition from animal rights movements and extremist organizations" (Shane 2005). At the same time, opposition to the uncorrectable cruelty of electrical "stunning," emphasized by United Poultry Concerns, PETA, and The Humane Society of the United States, has grown. In 2007, McDonald's claimed to be awaiting further developments, ConAgra Foods was urging suppliers to adopt the CAS system, and Burger King announced a purchasing preference for chicken meat "from processors who employ controlled atmosphere stunning (CAS) systems" (O'Keefe 2007).

By 2007, a plant called MBA Poultry in Tecumseh, Nebraska was using a carbon dioxide-based CAS system to "stun" broiler chickens, and four U.S. turkey plants had adopted the CO_2 system for male turkeys weighing around 40 pounds. One of these plants, Dakota Provisions in Huron, South Dakota, is run by Hutterite colonies (one of the Anabaptist faiths along with the Amish and Mennonites) and slaughters 4 million male turkeys a year (O'Keefe 2006).

What About Carbon Monoxide (CO)?

Some people wonder why carbon monoxide isn't used. Carbon monoxide is a colorless, odorless, painless, deadly gas. When I asked Mohan Raj about this he said, "CO can be used where available. For example, it was used in Belgium and Holland during 2003 avian influenza outbreaks. The problem is that a lethal concentration of this gas will affect all living creatures, not just humans who could be protected with breathing apparatus. Unlike argon or nitrogen, CO is explosive at 12.5 percent or higher and therefore imposes an additional safety burden" (Raj 2007).

According to the American Veterinary Medical Association's *2000 Report on Euthanasia*, carbon monoxide induces loss of consciousness without pain and with minimal discernable discomfort, but it is hazardous to humans precisely because it is highly toxic and difficult to detect. However, with properly designed ventilation systems, CO chambers, explosion-proof equipment, and commercial compression of the gas, CO can be used to kill dogs and cats (AVMA, 678–679). Nothing is said about birds, and it isn't clear why the problems presented by carbon monoxide could not be worked out in a poultry slaughter plant.

Destruction of Unwanted Chicks

What the video footage reveals is shocking: from the moment they're hatched, these turkeys are submerged into a world of misery. Dumped out of metal trays and jostled onto conveyor belts after being mechanically separated from cracked eggshells, the newly hatched turkeys are . . . sorted, sexed, debeaked, de-toed, and in some cases de-snooded. . . . Countless chicks become mangled from the machinery, are suffocated in plastic bags, or deemed "surplus" and dumped (along with injured chicks) into the same disposal system as the discarded eggshells they were separated from hours earlier.

—"Inside a Turkey Hatchery," 2007

Along with defective and slow-hatching female chicks, a quarter of a billion male chicks are trashed by the U.S. egg industry as soon as they hatch. Countless more millions of chickens, turkeys, ducks, and other birds born in hatcheries are treated the same. Upon breaking out of their shells in the mechanical incubators, instead of being sheltered by a mother hen's wings, the newborns are ground up alive, electrocuted, or thrown into trashcans where they slowly suffocate on top of one another, peeping to death as a human foot stomps them down to make room for more chicks.

A 1995 article in *Poultry World* gives a look at how baby chicks are tortured and killed in laboratories and commercial hatcheries around the world, using methods ranging from heat and electricity to carbon dioxide to garden waste cutters to a "superior method . . . with fast rotating knives" (Gerrits). The disadvantages of CO_2 from a humane standpoint are contrasted with the economic advantages: "The killing device is simple and affordable and the chicks stay unimpaired, so they can be sold to the pet-food industry especially for cat food."

Slaughter Culture: Submerged in a World of Misery

The treatment these birds receive is almost unbearable. They are tossed like potatoes, kicked if they happen to be dropped, and a report from one CFIA [Canadian Food Inspection Agency] vet tells of much more sadistic things. This is something that's surprised me—the sadisticness. I had always assumed much of their suffering was from neglect and an absentmindedness, but as we've been finding with our camera, people often have much more sinister intentions.

—Twyla Francois, Canada Head Inspector for Animals' Angels in an e-mail to the author, June 18, 2007

Using the turkeys as punching bags, ripping their heads off, and slaughtering conscious birds are among the accusations Mercy for Animals has filed against House of Raeford Farms, said to be the seventh largest turkey processing plant in the U.S.

—"Cruelty Charges Against Poultry Slaughterer," Farmed Animal Watch, 2007a

Workers were caught on video stomping on chickens, kicking them, and violently slamming them against floors and walls. Workers also ripped the animals' beaks off, twisted their heads off, spat tobacco into their eyes and mouths, spray-painted their faces, and squeezed their bodies so hard that the birds expelled feces—all while the chickens were still alive.

—Ward Harkavy, *Village Voice,* regarding PETA's undercover investigation of the Pilgrim's Pride chicken slaughter plant, in Moorefield, West Virginia, in 2004

Most of my fellow employees were extremely abusive to the chickens.

—Virgil Butler, former Tyson employee, in an affidavit signed June 30, 2003a

Abuse of chickens and turkeys by slaughterhouse workers is commonplace—everything from clubbing roosters to death at a Tyson plant because they didn't fit into the shackles, to breaking their legs to fit the shackles, to dumping lime on a truck full of live chickens left sitting on a loading dock to "keep the smell down" (Butler 2003d).

At another level, the cruelty is personal. Undercover tapes repeatedly show workers doing things to chickens and turkeys simply to vent their rage and exercise sadism: ripping birds' heads off, shoving dry ice up their rectums and watching them blow apart, stomping them to death, running over them with trucks, pulling eggs out of hens' bodies to throw at each other, and urinating on them (Butler 2003a; FAW 2007a; PETA).

In 2006 and 2007, workers at the U.K. turkey company Bernard Matthews were filmed using the turkeys as baseballs, hitting them with poles, and repeatedly kicking them. Accused of assaulting the birds, the workers claimed they were influenced by "peer pressure" and the "culture" at the slaughter plant (FAW 2007b).

Reduced to the status of objects, birds who, in the words of an undercover worker at a Perdue plant in Maryland, "scream[ed] so loud that you had to yell to the worker next to you, who was standing less than two feet away, just so he could hear you," are targets for extra-cruel treatment. The Perdue investigator wrote, for example: "Workers don't treat the animals aggressively only while they're hanging them. I saw an employee kick a chicken off the floor fan and routinely saw chickens being thrown around the room. Some birds manage to jump from the conveyer belt onto the floor before they can be shackled, so workers would grab them and throw them back toward the belt. A couple of times, the chickens were thrown so hard that the entire line would shake from the force of their bodies hitting the shackles" ("Working Undercover").

The Perdue investigator pointed out how invisible the chickens were in the training session: "The videos focused solely on worker safety and food contamination, not once instructing new employees as to the proper way to handle the animals. In fact, of the hours of videos shown, there was only about three seconds of footage of live animals, included in a montage of different ways workers might have to engage in lifting activity at various stations at a Perdue plant. Similarly, the presentations did not mention animal welfare, animal handling, or animal treatment at all" (Working Undercover").

Pain and Suffering in Birds

Chickens and turkeys—birds—experience pain, panic, fear, and distress the same as other animals including humans. Pain receptors, thermoreceptors, and physical-impact receptors responsive to noxious (tissue-damaging) stimuli have been identified in birds and characterized in chickens. Like mammals subjected to painful stimuli, chickens show a rapid increase in heart rate and blood pressure and behavioral changes consistent with those found in mammals indicating pain perception—

efforts to escape, distress cries, guarding of wounded body parts, and the passive immobility that develops in birds and other animals subjected to traumatic events that are aversive and that continue regardless of attempts by the victim to reduce or eliminate them (Gentle 1992).

Michael Gentle states in "Pain in Birds" that comparing the physiological responses of the nociceptors (pain receptors) found in chickens with those found in mammals, including humans, "It is clear that in terms of discharge patterns and receptive field size, they are very similar to those found in a variety of mammalian species." Birds, like mammals, he explains, have "a well developed sensory system to monitor very precisely external noxious or potentially noxious stimuli." He concludes, "Close similarity between birds and mammals in their physiological and behavioral responses to painful stimuli would argue for a comparable sensory and emotional experience" (Gentle 1992, 237–238, 243).

Birds are Intelligent Beings

In addition to comparable sensory and emotional experiences, birds have cognitive abilities "equivalent to those of mammals, even primates" (Rogers 1995, 217). This conclusion is shared by the Avian Brain Nomenclature Consortium, an international group of scientists. In a paper published in *Nature Neuroscience Reviews* in 2005, the Consortium presented the overwhelming evidence showing that a bird's brain is a highly complex organ of which fully 75 percent "is an intricately wired mass that processes information in much the same way as the vaunted human cerebral cortex." In light of this evidence, the Consortium is calling upon scientists around the world to adopt a new language to describe the various parts of the bird's brain in recognition of what is now known about avian intelligence, upsetting the "old system [that] stunted scientists' imaginations when it came to appreciating birds' brain power" (Weiss 2005). As for chickens in particular, scientists observe that "chickens evolved an impressive level of intelligence to help improve their survival" (Viegas).

The question is, what will we do with this knowledge?

"Humane Slaughter" Laws

There are no federal laws governing the handling and slaughter of poultry in the United States. Although birds represent 99 percent of land animals slaughtered for food in the U.S., they are excluded from the federal Humane Methods of Slaughter Act of 1958, amended in 1978, despite the fact that the eight bills presented at the 1957 hearings before the House Agriculture Subcommittee on Livestock and Feed Grains included both livestock and poultry. Those working to enact some form of humane slaughter legislation at the time relinquished poultry with the understanding that birds would subsequently be covered. Fifty years later, birds still are not covered.

Canada has an unenforceable Recommended Code of Practice, and the United Kingdom leaves enforcement of its welfare laws to the Meat Hygiene Service and to an unenforceable Ministry of Agricultural Code of Practice. In the European Union, a 1997 draft proposal by the European Commission to strengthen an EU Directive to set legally binding stunning standards for birds and mammals "seems to have disappeared" (Stevenson 2001, 2.2).

As already noted, the Poultry Products Inspection Regulations of the U.S. Department of Agriculture only states, "Poultry shall be slaughtered in accordance with good commercial practices in a manner that will result in thorough bleeding of the carcasses and assure that breathing has stopped prior to scalding" (PPIA).

The concern about breathing is not humanitarian but to prevent condemnations resulting from "redskins" produced when live birds enter the scald tank, and to prevent live birds from inhaling the contaminated scald water into which they are plunged after bleeding from the neck while hanging upside down in the bleedout tunnel, where an unspecified number of birds drown in pools of blood when the conveyer belt dips too close to the floor.

Redskins, which the Department of Agriculture calls "cadavers," as opposed to properly bled "carcasses," are permitted in one or two birds per 1,000 slaughtered, according to USDA guidelines (Brant). However, former Tyson employee Virgil Butler said that of the 80,000 chickens killed every night at the Arkansas plant where he worked, about one in every three were still alive when they entered the scald tank (Singer and Mason 2006, 26–27). Butler states:

> I was responsible for trying to slit the throats of the chickens the machine missed on the nights I worked the killing room. Our line runs 182 shackles [shackled birds] per minute. It is physically impossible to check them all. Therefore, they are scalded alive. When this happens, the chickens flop, scream, kick, and their eyeballs pop out of their heads. They often come out of the other end with broken bones and disfigured and missing body parts because they've struggled so much in the tank (2003a).

Efforts to Obtain "Humane" Poultry Slaughter Coverage in the United States

California

Animal protectionists in the United States have supported legislation to extend humane slaughter coverage to poultry. California initiated legislation in 1991 by adding an amendment to California's Methods of Slaughter law requiring that animals killed for food must be rendered insensible to pain before slaughter. However, small birds and laying hens are excluded, and in 1994 the California Department of Food and Agriculture arbitrarily invited ritual slaughterers to apply for an exemption to the regulations.

The Humane Slaughter of Poultry Regulations, Title 3 CA Code of Regulations, Article 15.1, Section 1245.1–1245.16 went into effect on December 14, 1996. It codifies practices listed under the federal Humane Methods of Slaughter Act (which excludes birds), with the exception that whereas mammals are to be rendered "insensible to pain" before being shackled, birds may be rendered "insensible to pain" after being shackled. In 2005, a bill was introduced (SB 662) that would have put spent laying hens and small game birds under California's Methods of Slaughter law, but it was not enacted. The California Animal Association, of which I was a member, sought in vain for a single poultry scientist in the United States willing to speak publicly on behalf of the bill.

The Federal Level

In the 1990s, three bills were introduced in the U.S. House of Representatives by Congressman Andrew Jacobs of Indiana—H.R. 4124 (1992), H.R. 649 (1993), and H.R. 264 (1995). They sought to amend the 1957 Poultry Products Inspection Act to provide for the "humane" slaughter of poultry similar to how the 1906 Meat Inspection Act was used as a basis for determining the coverage of "cattle, sheep, swine, goats, horses, mules, and other equines" under the 1958 Humane Methods of Slaughter Act. However, all of these bills died in the House Agricultural Livestock Subcommittee to which they were referred.

On September 28, 1994, House Subcommittee chairman Harold Volkmer, of Missouri, held a hearing on H.R. 649, the "Humane Methods of Poultry Slaughter Act." United Poultry Concerns, the Animal Legal Defense Fund, and the Animal Welfare Institute presented oral and written testimony on behalf of this bill. However, Representative Volkmer announced his opposition to the bill at the outset and made jokes about killing chickens on the farm during the hearing.

Frustrated by government, several animal protection organizations petitioned the U.S. Department of Agriculture on November 21, 1995, to use its statutory authority to extend humane slaughter protection to poultry through an amendment of the poultry products inspection regulations issued under the Poultry Products Inspection Act. *Petition for Rulemaking Regarding Regulations Issued Under the Poultry Products Inspection Act (PPIA), 21, U.S.C. Sec. 451, et seq.* was submitted to the Department of Agriculture on November 21, 1995, by the Animal Legal Defense Fund, the Animal Welfare Institute, and the Society for Animal Protective Legislation, AWI's legislative arm. If the petition had been granted, USDA inspection regulations for poultry would have included a provision for the "humane" slaughter of poultry, similar to inspection regulations covering the "humane" slaughter of most mammals killed for food in the United States. The petition wasn't granted.

Ten years later to the day, in light of investigations documenting extreme cruelty to birds in slaughter plants in Arkansas, West Virginia, and Maryland, The Humane Society of the United States and the California-based East Bay Animal Advocates sued the U.S. Department of Agriculture on November 21, 2005. The suit, filed in federal district court in San Francisco, challenged the exclusion of poultry from the Humane Methods of Slaughter Act and sought to ensure that birds are unconscious before being slaughtered. The lawsuit stated that poultry plants hang live birds injuriously in metal shackles and subject them to paralyzing electric shocks before cutting their necks and dumping them into tanks of scalding, feces-contaminated water while they are still alive (Evans and Page; Williamson).

However, in 2008, U.S. District Court judge Marilyn Hall Patel dismissed the lawsuit, stating that the language of previous bills indicated that Congress "intended to exclude poultry from the definition of livestock when it enacted H.R. 8308," the bill that became the Humane Methods of Slaughter Act in 1958 (FAW 2008).

For an idea of what birds are up against in the United States, in 2005 the USDA's Food Safety and Inspection Service (FSIS) published a notice in the *Federal Register* titled "Treatment of Live Poultry Before Slaughter" (USDA-FSIS). Noting that it had received over 20,000 letters from members of the public concerned about humane treatment of livestock, including 13,000 e-mails supporting legislation to include provisions for the humane treatment of poultry under the Humane Methods of Slaughter Act, the agency invited public comment. United

Poultry Concerns submitted comment on October 25 (UPC 2005). On June 25, 2007, I phoned FSIS for a status report on the notice and was told by the Director of Regulatory Staff, Rachel Edelstein, that it was "just a reminder." There was, and is, no plan to change existing regulations, she said.

It is inexcusable that the huge majority of land animals slaughtered for human consumption in this country are denied "humane slaughter" coverage. Cruelty prosecutions are impossible under these circumstances. The U.S. government and the poultry industry have no accountability regarding their treatment of the billions of birds they handle and kill, while the birds are alive and capable of experiencing what is being done to them. The effort to extend coverage should not be regarded as a sanction for slaughter or a salve for conscience. Rather, the absence of a law conveys the false notion to the general public, and to those who work directly with poultry, that these birds do not suffer, or that their suffering does not matter, and that humans have no merciful obligation to them even to the nominal extent granted to cattle, sheep, and pigs.

At the same time, to those who say that vegetarianism will not come overnight, it can be said with even greater assurance that "humane slaughter" will never come at all, because the slaughter process is inherently inhumane, and the slaughter of the innocent is wrong, and because the poultry industry, even in countries where humane slaughter laws may exist, is, for all practical purposes, ungovernable. Humane slaughter is an illusion. Rendering the slaughter process less inhumane is a possibility. A question is whether "humane slaughter" legislation for poultry will speed or delay the day when regarding a fellow creature as food is no longer an option.

CHAPTER 6

A NEW BEGINNING

With increased knowledge of the behaviour and cognitive abilities of the chicken has come the realization that the chicken is not an inferior species to be treated merely as a food source.

—Lesley J. Rogers, *The Development of Brain and Behaviour in the Chicken*, 213

The plea for ethical veganism, which rejects the treatment of birds and other animals as a food source or other commodity, is sometimes mistaken as a plea for dietary purity and elitism, as if formalistic food exercises and barren piety were the point of the desire to get the slaughterhouse out of one's kitchen and out of one's system. Abstractions such as "vegetarian" and "vegan" mask the experiential and philosophical roots of a plant-based diet. They make the realities of food animal production and consumption seem abstract and trivial, mere matters of ideological preference and consequence, or of individual taste, like selecting a shirt or hair color.

However, the decision that has led millions of people to stop eating other animals is not rooted in arid adherence to diet or dogma but in the desire to eliminate the kinds of experiences that using animals for food confers upon beings with feelings. The philosophic vegetarian believes with Isaac Bashevis Singer that even if God or Nature sides with the killers, one is obliged to protest. The human commitment to harmony, justice, peace, and love is ironic as long as we continue to support the suffering and shame of the slaughterhouse and its satellite operations.

Vegetarians do not eat animals, but, according to the traditional use of the term, they may choose to consume dairy products and eggs, in which case they are called lacto-ovo (milk and egg) vegetarians. In reality, the distinction between meat on the one hand and dairy products and eggs on the other is moot, as the production of milk and eggs involves as much cruelty and killing as meat production does: surplus cockerels and calves, as well as spent hens and cows, have been slaughtered, bludgeoned, drowned, ditched, and buried alive through the ages. Spent commercial dairy cows and laying hens endure agonizing days of preslaughter starvation and long trips to the slaughterhouse because of their low market value.

In reality, milk and eggs are as much a part of an animal as meat is. No less than muscles, these parts comprise the physiological, metabolic, and hormonal activities of an animal's body. A hen's egg is a generative cell, or ovum, with a store of food and immunity for an embryo that, in nature, would normally be growing inside the egg. Milk is the provision of food and immunity produced by the body of a female mammal for her nursing offspring.

Historically, ethical vegetarianism has rejected the eating of an animal's body as this requires killing the animal specifically for the purpose of consumption. The ethical vegetarian regards killing an unoffending creature simply to satisfy one's palate and to fit into society with revulsion. Premeditating the premature death of an animal is also disdained. Plutarch mourned, "But for the sake of some little mouthful of flesh we deprive a soul of the sun and light, and of that proportion of life and time it had been born into the world to enjoy."

Many people believe that the degradation of other creatures is an inherent feature of using them for food. The animals are dominated by humans. Their food is chosen; their social, familial, and physical environment is controlled; their reproductive organs and activities are manipulated; and the length of their lives is determined by humans. They can be abused at will based on economic "necessity."

In nature, animals exist for their own reasons, not only for others' use. In production agriculture, animals are brought into the world solely to be used. Any happiness they may enjoy is secondary to their utility and dependent upon the "permission" of their owner, who has complete jurisdiction over their lives, including the right to kill them at any time. Those who desire to end this arrangement, but feel they must detach themselves from it by stages, should begin by eliminating dairy products, eggs, poultry, and fish from their diet, because the cruelty embodied in animal products is compounded by the number and longevity of animals used to produce these particular products.

Only consider how many people now eat two or three chicken breasts at a single meal, or several wings as a snack, or the fact that dairy cows and battery hens are tormented not only for months but for a year, or years, before being slaughtered. The countless numbers of fish killed each year for human consumption are not even measured in terms of individuals but in terms of metric tons. Fish are increasingly being raised in the factory farm systems known as aquaculture as a result of human overpopulation and—ironically, given the filth of industrialized fish farming—water pollution. Fish are being subjected to genetic engineering, forced rapid growth, drugs, and diseases of confinement, making them, in the most ultimately gruesome sense, "the chickens of the sea."

Free-Range

The U.S. Department of Agriculture states in an information bulletin: "Producers must demonstrate to the agency that poultry has been allowed access to the outside." Hypothetically, the door to the coop could be open for five minutes, and the chicken could still qualify as free-range. Also, labels are not necessarily verified.

—Samantha Storey, "Know Your Meat," 2004.

"Free-Range" Birds Kept for Meat

A growing number of people are looking to "free-range" as an alternative to factory farm products. "Free-range" conveys a positive image of drug-free animals living outdoors, as nature intended. Historically, the term "range" meant that, in addition to living outside and getting exercise, the animals could sustain themselves on the land they occupied. However, this is no longer the case.

Birds raised for meat in the United States may be sold as "free-range" or "free roaming" if they have USDA-certified access to the outdoors, a requirement that isn't monitored. No other criteria—vegetation, range size, number of birds, or space per bird—are defined by the Food Safety and Inspection Service of the U.S. Department of Agriculture, which reviews and approves labels for federally inspected meat products. A USDA staffer told me, "Places I've visited may have just a gravel yard with no alfalfa or other vegetation. The birds can exercise but cannot range—that is, sustain themselves" (Ricker).

The USDA does not require that the "range" be kept fresh. The area may be nothing but a mud yard saturated with droppings and intestinal coccidia and other parasites. Chickens spend much of their time close to the house, scratching, dustbathing, and wearing away the grass. A static house and pasture becomes unsanitary when hundreds or thousands of birds are collected in a small area. For this reason, a system of pasture rotation made up of small movable houses holding 300 or fewer birds with constant access to fresh grass is needed.

Yet even a pasture arrangement may be of poor quality, as a visitor to Polyface Farm in the Shenandoah Valley of Virginia, often hailed as a model enterprise, discovered. Mary Finelli wrote: "I toured Polyface on a sweltering day. Chickens were in tiny cages with tin roofs in the beating sun, panting like mad. The cages were located over manure piles which the birds were supposed to eat larvae from. Rabbits were kept in factory-farm conditions in suspended, barren, wire cages."

"Free-range" turkeys may fare no better. Terry Cummings of Poplar Spring Animal Sanctuary in Poolesville, Maryland, described the treatment of turkeys reported by two visitors to Springfield Farm in Sparks, Maryland, on a frigid December day: "The turkeys were housed in an open field in the freezing cold with no shelter except a small, wooden, tarp-covered structure, which was only big enough for half of the turkeys. The others were huddled together shivering in the weather, exposed to the elements. The farmer roughly grabbed the turkeys by their legs and held them upside down while they flapped desperately to right themselves. This is how he carried them."

East Bay Animal Advocates uncovered horrible conditions at Diestel Turkey Ranch, a "free-range" farm in Northern California and supplier to the Whole Foods Market chain, which in 2005 announced an initiative called "Farm Animal Compassionate Standards." Investigated from 2004 to 2005, the farm was found to be indistinguishable from a standard industrial turkey operation. It was an Orwellian-style *Animal Farm* posing as its opposite—complete with debeaked young turkeys crowded miserably in filthy buildings and not a turkey to be seen outside (Morrissey).

"Free-Range" Hens Kept for Eggs

Eggs sold in the United States may be falsely advertised as "range," because there is no definition of husbandry terms regulating the sale of eggs in this country. The National Supervisor of Shell Eggs at the U.S. Department of Agriculture merely administers a voluntary program in which producers can use the USDA grade mark if the eggs have been packaged under USDA supervision.

An example of what may pass as free-range or cage-free is a brand of eggs marketed by Pleasant View Egg Farm in Winfield, Pennsylvania, as "happy hen Organic Fertile Brown Eggs." The hens I saw in the early 1990s were advertised as "free running in a natural setting [and] humanely housed in healthy, open-sided housing, for daily sunning—something Happy Hens really enjoy."

On June 24, 1992, Joe Moyer, the owner, provided a tour. Three Happy Hen houses were perched on remote Amish contract farms in Logantown, Pennsylvania. "Humanely-housed" inside the long barns were 6,800 hens and one rooster for every 100 hens. As we drove up to one of the houses on a beautiful June day, we saw a crowd of chicken faces pressed against the netting of a building in the middle of grass and woods they never set foot in. Inside, the birds were wall to wall. They were severely debeaked. Their feathers were straggly, drab, and worn away. When we commented on the condition of the hens' feathers, the owner bragged, "We have a saying: 'The rougher they look, the better they lay.'"

Chickens can live active lives for 15 years or longer, but after a year or two, commercial free-range hens are hauled away the same as battery hens. Noncommercial family farms generally keep their "girls" two or three more years before replacing the entire flock. Many spent free-range hens go to the highest bidder, usually a live poultry market. The Happy Hen hens were trucked to live poultry markets in Pennsylvania, New Jersey, and New York City where they fetched a dollar per bird, compared to 2 to 25 cents per bird from a spent-fowl slaughter plant.

Even if free-range eggs were the humane alternative people would like to think they are, the problem of excess roosters would remain. Apart from the males used to replenish flocks or to fertilize eggs for niche markets, roosters have no value in egg production. Therefore, the brothers of the free-range hens—like the brothers of the battery-caged hens—are trashed at birth and sold to laboratories, "Easter chick" peddlers, and pet stores. No amount of advertising changes this fact.

Organic Standards

"Organic" refers to the ingredients that are fed to animals raised for food. Under the National Organic Program, created by Congress under the Organic Foods Production Act of 1990 and administered by the U.S. Department of Agriculture since 2002, the USDA is charged with inspecting the source of food and certifying that meat, poultry, eggs, and dairy products labeled "organic" come from animals who were fed no antibiotics, growth hormones, or animal by-products. Synthetic pesticides and fertilizers, bioengineering, and ionizing radiation are likewise prohibited (Kuepper).

The USDA does not verify organic products. It accredits organic certifiers to authorize the Certified Organic or USDA organic labels.

In 2002, the National Organic Standards Board withstood intense lobbying by the poultry industry by voting to require outdoor access for poultry. The board was empowered by Congress under the Organic Foods Production Act to make recommendations to the USDA regarding the contents of the Organic Rule, which defines substances and practices that can be used to produce products with the USDA organic label. The 15-member board voted 12 to 1 to keep the outdoor access for poultry rule (HSUS 2002).

In 2006, the National Institute for Animal Agriculture, an industry trade group, lobbied the USDA's Animal and Plant Health Inspection Service to change the National Organic Program regulations to eliminate the words "access to the outdoors" as a requirement for the production of USDA-certified organic poultry (NIAA, 22), citing avian influenza as the reason. Currently, the worldwide poultry industry is urging governments to require enclosure for all birds raised for food (GRAIN). This is already having an effect on organic producers and easing the way for industrialized companies to move in.

Tyson Foods, for example, has entered the organic market with a subsidiary brand of chicken based on organic feed ingredients and claims that Tyson's "organic" chickens have "room to spread their wings." What this shows, says Steve Striffler in his book *Chicken*, is how easy it is "to produce 'healthy' organic and free-range chicken in a way that differs very little from industrial chicken" and "why companies such as Tyson have moved so quickly into organic chicken. It requires very little change in the way that industry leaders do business" (156, 168).

Another example of what organic may actually mean can be found in Peter Singer and Jim Mason's book *The Way We Eat: Why Our Food Choices Matter*. As part of their research, the authors visited Pete and Gerry's Organic Eggs, an organic egg farm in New Hampshire where 100,000 debeaked hens were housed in six long sheds. What they saw in September was "a shock."

> The shed was about 60 feet wide and 400 feet long. Covering the floor, stretching away into the distance, was a sea of brown hens, so crowded that the shed floor was visible only down the center of the shed where the hens had left a gap in between the feed and water areas on each side. There were, Jesse [the manager] told us, about 20,000 birds in that shed. Each of them had 1.2 square feet, or 173 square inches, of space. . . .
>
> The most controversial [organic] rule, Jesse told us, is the requirement for outdoor access for the hens. . . .

> "So these hens have outdoor access?" we asked.
>
> Jesse pointed to a bare patch of dirt between the shed we had been in and a neighboring shed. "There are penned in areas over there, and around the back," he said. But the shed we had been in didn't seem to allow any way of getting outside. We asked Jesse how the birds got out. "It's sealed," he acknowledged. "I sealed it up about three or four weeks ago, because of the time of year. The USDA has exceptions for that, depending on the climate... But the controversy is really about disease [avian influenza], if the hens come into contact with wild birds."
>
> . . . . We pressed on: "If we had come a month ago, and it was a warm day, would there be hens out there?"
>
> "If it was a clear day, and we could be sure that there were no wild birds flying over, yeah."
>
> "But you can never be sure that there would be no wild birds flying over!"
>
> "Right. There's the rub. But we've had the doors open on some days. Not many birds go out [to the patch of dirt]."
>
> "And you don't get problems with the inspectors about that?" we asked.
>
> "No, nobody is getting problems about that, at this point" (Singer and Mason 2006, 101–105).

While a well-managed organic or free-range system is ethically and ecologically superior to the battery-cage system, it does not solve the problem of the unnatural isolation of the birds from other sexes and age groups of their species and from other species. Chickens enjoy the company of other creatures and get along well with them. They exchange benefits and express the vitalities, intelligence, and charms of their nature.

Veterinarian Holley Cheever describes an arrangement between Rafe, her horse, and her Rhode Island Red hen, Aurora: "She has learned that an excellent source of flies in the summertime is our horse's belly, where flies love to cluster and feed. She squats directly beneath him, waiting for a fly to land, and hops up to snatch it with unerring accuracy. Not only does she never miss her target, she doesn't even touch Rafe's sensitive abdomen, which would prompt him to kick up at his underbelly, no doubt injuring her in the process. There is something particularly amusing about the stance she assumes before the strike which reminds me of a pro basketball player's body English as he goes for a rebound."

See for Yourself

Only oppressors can deny the importance of suffering to the individuals who suffer or who respond to that suffering.

—Carol J. Adams and Marjorie Procter-Smith,
"Taking Life or 'Taking on Life,'" 305

People should visit as many farmed animal systems as possible to see for themselves what goes on and ask questions. When visiting a battery-caged hen complex, for instance, they should ask to see the "old" or "spent" hens who have been locked up for seven or eight months, churning out eggs. Otherwise the management will show you only the newly installed hens who have not yet lost most of their feathers and become covered with raw sores.

When you visit, look at the faces and eyes of the animals and observe their body language. Notice their voices. The idea that human beings cannot logically recognize suffering in a chicken, or draw meaningful conclusions about how a human being would react to the conditions under which a caged hen lives, is ridiculous. There is a basis for empathy and understanding in the fact that human evolutionary continuity with other creatures enables us to recognize and infer, in those creatures, experiences similar to our own. The fact that animals are forcibly confined in environments that reflect human nature, not theirs, means that they are suffering much more than we know and in ways that we cannot fathom. If they preferred being packed together without contact with the outside world, then we would not need intensive physical confinement facilities, and mutilations such as debeaking, since they would voluntarily come together, live cordially, and save us money. The egg industry thinks nothing of claiming that a mutilated hen in a cage is "happy," "content," and "singing," yet will turn around and try to intimidate you with accusations of "anthropomorphism" if you logically insist that the hen is miserable.

Veterinarian Michael W. Fox states that we cannot argue that the more domesticated an animal is, the less freedom the animal needs, because domestication is more a change in the relationship between the animal and the environment than within the animal. If this were not the case, he says, "then highly domesticated dogs, pigs, and battery-caged laying hens would not have the capacity to become feral and, when given the opportunity, express the full range of behaviors that their wild counterparts possess. Domesticated animals can differ greatly in size and shape from their wild relatives, but they differ little in terms of their behavioral repertoires" (Fox, 205–206).

One method that we use to assuage our guilt about the way we treat farmed animals is to plead that, having no basis for comparison, they cannot know that their lives are desolate. We might as well use this plea to absolve ourselves of responsibility toward anyone in the world who, we decide, because they have never known anything but misery, "cannot know that their lives are desolate." Are we prepared to say that babies who are born junkies as a result of their mother's drug habit, having no basis for comparison, cannot know that their lives are desolate? Is this an acceptable argument for children born into slavery? Even if this were so, where does such thinking constructively lead?

As a matter of fact, animals do "know," because knowing is an organic process far deeper than words and concepts can express. Every bodily cell is a repository of experiences including memory and expectation as elements of a particular moment in the life of that particular cell. The look in a creature's eyes tells us a whole lot about what he or she "knows." Freedom and well-being, as Michael Fox observes, "are more than intellectual concepts. They are a subjective aspect of being, not exclusive to humanity, but inclusive of all life. This is not an anthropomorphic claim. It is logically probable and empirically verifiable" (208).

It is remarkable how far a person may go to justify the exploitation of other creatures. An example is the poultry researcher who said that if he thought animals had a concept of the past and the future, he would be a vegetarian (Duncan 1993). He proposed that the ability to conceptualize the past and the future should be accepted as the morally relevant difference between other living beings (their intrinsic and ultimate worth) and ourselves. In fact, we do not know that other animals lack this ability; we merely assume (and perhaps hope) that they do. But if they do, so what? Are we justified in saying, "You lack the ability to form a concept of the past and the future; therefore I have a right to kill you"?

As for humans, we have, at best, an ambiguous relationship to the past and the future, both morally and conceptually. As well as seeing some things clearly, human beings constantly reinvent the past and fabricate in their minds an idealized future, both consciously and unconsciously, for good and for evil. There is excellent evidence that we humans have a very poor memory of our own past conduct. How many millions—billions—of people have been murdered in the name of the "past" and the "future"? Even if other kinds of animals do experience the past and the future differently than we do, their activities, and their responses to new as well as to familiar situations, show that, like us, they anticipate, plan, and remember. It is precisely on such grounds that an avian scientist has condemned the battery-cage system for laying hens, charging, "In no way can these living conditions meet the demands of a complex nervous system designed to form a multitude of memories and to make complex decisions" (Rogers 1995, 218).

If the ability to remember is a basis for determining a creature's right to be or not to be, then what do we do about all those people (including ourselves) whose sense of their own past excludes a vicarious identification with their own victims? Carol J. Adams and Marjorie Procter-Smith said, "The voice of the voiceless offers a truth that the voice of the expert can never offer: it offers the memory of suffering and the truths of subjugated knowledge" (302). This memory and these truths comprehend a totally different past from the past of the experts.

Cruel Research

[W]e're a long, long way from using up genetic gain potential in rate of growth, feed conversion, and how we shape our birds. . . . More and more, the biochemist, physiologist, virologist, immunologist, microbiologist, and nutritionist will complement the geneticist's activities in poultry breeding programs. . . . All of this translates into relatively large investments in research and development, which in turn must be justified on an R.O.I. [return on investment] potential basis as it relates to the marketplace—market size, and market penetration.

—Wentworth Hubbard, President of Hubbard Farms, a subsidiary of Merck & Co. since 1974

Brazilian poultry producers go through a marathon in the search of new technologies. Acclimatized chicken sheds, automatic feed distribution, and vaccination in eggs are part of the routine in poultry farms. With research they have also already obtained a chicken that grows very fast. The great question now is how to breed a bird with a physiological system that keeps up with this growth.

—Isaura Daniel, "The Race for the Perfect Chicken"

Chickens are sensitive living beings. To the companies that own them, they are commodities and investments no different from the Byzantine paraphernalia that is used to manipulate and kill them. While modern genetic, chemical, and management practices have combined to create costly and painful diseases in chickens, these diseases generate huge sums of money. They are the testing ground and target for expensive new pharmaceuticals. Lucrative, perennially renewable government, state university, and private sector contracts are involved. Companies such as Tyson, Merck, and Bayer have their own poultry research facilities. We hear little about these facilities, as the research is kept quiet and is seldom published.

Until 1988, Eldon Kienholz (1928–1993) was a full professor, specializing in poultry nutrition in the Department of Animal Sciences at Colorado State University. In 1988, rather than continue to perform the cruel experiments on chickens and turkeys his tenure required, he chose to retire and speak out. In an interview with me, Dr. Kienholz talked about some of his research projects (Davis 1991).

Could you give an example of the kind of research you did?
Yes. I knew that wings and tails of birds were unnecessary to the commercial production of poultry meat, so I did research to show that a grower could save about 15 percent of feed costs by cutting off the tails and wings of broiler chicks and turkey poults soon after hatching. I gave papers on that at national meetings and attracted a great deal of interest.

What caused you to become skeptical about your work? Was it a utilitarian consideration? A moral twinge?
A moral twinge. Somehow it didn't feel right to be cutting off the wings of newly hatched birds. Later, some of them couldn't get up onto their feet when they fell over. It wasn't pleasant seeing them spin around on their side trying to get back onto their feet, without their wings.

The decision to consume animal products involves one morally with millions of animals beyond those being used strictly in food production. Huge numbers of chickens and other farmed animals are subjected to painful and degrading experiments on behalf of the food industry every single day (Davis 2003b). Their status as flocks and herds ensures that vast numbers will be used in agricultural experiments simulating commercial production situations.

In 1988, the *Guide for the Care and Use of Agricultural Animals in Agricultural Research and Teaching* was published by a consortium executive committee, based on the 1985 edition of the *Guide for the Care and Use of Laboratory Animals* published by the National Institutes of Health (NIH). It was developed in response to policies of the U.S. Public Health Service's Office for Protection from Research Risks that went into effect in January 1986, which resulted in the inclusion of agricultural animals used in agricultural research and teaching in many institutions' animal care and use programs for the first time (CEC). It was revised (so to speak) in 1997.

The equivocation of the *Guide* is evident in the fact that while professing to encourage scientists to continue seeking improved methods of farmed-animal care and use, the authors "accept" procedures that may cause "some temporary discomfort or pain" if these are standard commercial practices "warranted in context of agricultural production" (CEC, 5). This leaves the door wide open; the proviso that painful and otherwise distressful experiments should be "performed with precautions taken to reduce pain, stress, and infection" is undermined by the fact that normal agricultural experiments on live birds and other farmed animals are either deliberately designed to produce pain, stress, fear, and infection, or else they cannot be performed without producing these conditions.

An insolent cruelty operates. A researcher told me that shaving hens naked with sheep shears in heat-stress studies was "very humane . . . just like a haircut" (Coon, 1994). He and his colleague joked in a published article: "The sheared areas of the hens had no feather coverage; however, they tended to have a white, fuzzy look because the stub for the feather quill remained with the skin" (Peguri and Coon, 1319). Similarly, industry people call the featherless chickens they're producing in laboratories "ugly little beasts" and "ugly birds" (Young 2002; Van der Sluis 2007).

Those wanting an idea of the kinds of experiments that are regularly conducted on chickens, turkeys, ducks, quails, and other domesticated fowl should consult the pages of publications such as *Poultry Science* and *International Hatchery Practice* and visit the websites www.worldpoultry.net and www.thepoultrysite.com. The USDA bibliographical series publication "Stress in Poultry: January 1979–August 1990" shows the nadir that can be reached by researchers and others as a result of society's detachment from the fate of farmed animals, aided by our unquestioning acceptance of what we have been taught to revere as "science." Moreover, countless atrocious experiments are conducted that are never even published.

To illustrate the kinds of things that are done to hens in the name of egg production, let us look at three examples. In experiments published in *Poultry Science* in 1984, researchers manually inserted inflated balloons, shell membranes, and tampons into the uteri of hens and gave them anti-inflammatory and immunosuppressive drugs to determine possible causes of shell-less eggs (Roland, et al.). The presence of the foreign shell membranes and tampons in the hens' uteri caused high fever, vomiting, diarrhea, and death.

"The balloons were the most difficult of all the materials to insert into the uteri," the researchers wrote. "If a hen had not accepted the balloon after it had been inserted into the uterus 3 or 4 times, the bird was not used. A punctured uterus was believed to be the cause of death of one hen in this group. In most cases the materials remained in the uteri from 1 to 48 hr. In one instance, a Rely tampon was expelled within 1 hr, but the bird still died. In other cases, tampons or balloons remained in the uteri overnight and were sometimes enclosed in shell membranes. Some treated [noncontrol] hens died within 8 hr; however, most hens died between 14 to 18 hr after insertion." The researchers concluded that the hen's reproductive system might serve as a model for studying human toxic shock syndrome.

In a study published in the *Journal of Applied Poultry Research*, a researcher at Purdue University published an experiment titled "Effect of Red Plastic Lenses on Egg Production, Feed per Dozen Eggs, and Mortality of Laying Hens." The hens' eyes were fitted with red contact lenses. "Seven hundred and ninety Dekalk L pullets, 10 weeks of age, were obtained from a local hatchery. Beaks of all pullets [the young hens] had been trimmed at day one by using a hot blade; non-trimmed pullets were not available at that time," the researcher wrote (Adams 1992).

At 17 weeks old, the hens were moved from the caged-pullet house to the caged-layer house. According to the researcher, "Considerable mortality occurred among birds with lenses between 2 and 8 weeks after moving the birds to the layer facility. Mortality was attributed to an inability of the birds to find the feed." He concluded that a lens was being developed by a company called Animalens, which claimed that the use of contact lenses contributed to reduced feed usage by decreasing feed consumption, bird activity, or both. Animalens funded the research on what company president Randy Wise kiddingly referred to in a phone conversation with me, on February 10, 1992, as "chicken goggles."

In 1991, I undertook to investigate the use of permanent red contact lenses in laying hens after receiving two written complaints from employees in the poultry unit at California Polytechnic State University, in San Luis Obispo. United Poultry Concerns' report *Red Contact Lenses for Chickens: A Benighted Concept* cites employees' charge that a lens experiment on caged hens sponsored by Animalens was causing severe eye infections, abnormal behavior, and blindness, and preventing the hens from closing their eyes normally because the lenses were so large. The hens were "pecking the air" and "rubbing their eyes repeatedly on their wings." The Animalens trainer sent by the company to insert the lenses hadn't even bothered to wash her hands.

An article in the school newspaper noted that the lenses—made of cheap, non-gas-permeable plastic that prevented the hens' eyes from breathing—had caused severe infections that worsened with time, and that those working with the hens were discouraged from helping them, in keeping with commercial practice. According to a student, "It's just not profitable to spend time treating the infections. What's taught in class is the less time you handle these birds, the more money you'll make" (Bock).

The photograph that accompanied the article showed a severely debeaked hen whose left eye appeared to have dissolved under the lens. The hens received no veterinary care or treatment during or after the experiment. They developed painful corneal ulcers and blindness and were left to languish with the lenses in their eyes for months in the poultry unit. United Poultry Concerns and a local organization, Action for Animals' Rights, sought in vain to persuade the university to place the remaining hens, at our expense, with veterinary ophthalmologist Nedim Buyukmihci, who offered them a permanent home. When we succeeded in covertly removing several hens from the poultry unit on July 25, 1991, they were taken to Dr. Buyukmihci, who immediately removed the lenses from the hens' eyes with fine eye forceps and photographed the injuries.

Research to increase egg production and "welfare" by destroying hens' ability to see continues. A professor of agricultural food and community ethics at Michigan State University, Paul Thompson, calls the breeding of blind chickens for egg production "emblematic" of the "ethical conundrum" involved in adjusting "the animal, rather than adjusting the production system." Since (he claims) blind chickens "don't mind being crowded together so much as normal chickens do," what most people would consider horrible really isn't: "If you think it's the welfare of the individual animal that really matters here, how the animals are doing, then it would be more humane to have these blind chickens" (Kestenbaum).

Researchers at the University of Guelph in Ontario, Canada claim that blind chickens do in fact lay more eggs (Dickenson). They're investigating how light affects egg production patterns in a strain of White Leghorn hens they call "blind Smokey Joes," with a view to giving egg producers "more tools to alter light techniques for higher performance." Total darkness may turn out to be better for the economics of egg production than any light at all for laying hens. It can then be promoted as "reduced susceptibility to stress" that increases profitability as well.

No Protection

These chickens have been morally abandoned in a society and a world in which "food" animals have historically been abandoned. Farmed animals are excluded from the definition of "animal" under the federal Laboratory Animal Welfare Act of the United States, which was created in 1966 to protect certain warm-blooded animals used in nonagricultural—biomedical and basic—research. Animals used in agricultural research are not covered by the Animal Welfare Act (AWA).

Apart from certain wild categories, birds are excluded from any form of federal protection in the United States. In 1994, the U.S. Department of Agriculture successfully appealed to the U.S. Court of Appeals for the District of Columbia Circuit to abstain from regulating birds, rats, and mice under the Animal Welfare Act. The 1992 case against the USDA for refusing to include these animals was dismissed for lack of standing of the plaintiffs, the Animal Legal Defense Fund and The Humane Society of the United States. Denied standing, the plaintiffs could not represent the interests of birds, rats, and mice before the court *(Petition)*.

In 2007, in response to ongoing pressure by the American Anti-Vivisection Society, the USDA's Animal and Plant Health Inspection Service (APHIS) drafted a proposed rule reflecting the Department of Agriculture's intention to regulate the pet bird trade (Letterman). Birds used for research and agricultural purposes will not be affected. The American Anti-Vivisection Society's overview of the Animal Welfare Act can be read at www.aavs.org/campaign_awahistory.htm.

While every effort must be made to extend legal protection to farmed animals, the question remains as to how far the law can protect a creature whom it has defined in advance as a piece of property, a thing without rights, over which the lawmakers have appointed themselves absolute jurisdiction. A wag asks what difference it makes how chickens are raised, since they are going to end up on a plate anyway. "I pointed this out to the waiter. He said, 'All of our chicken is free-range.' And I said, 'He doesn't look very *free* there on that *plate*'" (Briggs).

Giving chickens a decent life before killing them is met with serious ridicule. This is a clear reason to stop raising them for food. It is not their inevitable death that seems to justify our abuse of them when they are alive. Since death is the fate we all share, we do not generalize the argument. The justification is that we are deliberately going to kill them.

There is a felt inconsistency in valuing a fellow creature so little and yet insisting that he or she be granted a semblance of tolerable existence prior to execution. So wanton can our disrespect for our victims become under these circumstances that any churlish sentiment or behavior seems fit to exercise. It is contemptible to assert that humans have no responsibility or that it makes no sense to comfort the life of a being brought into the world simply to die for us. The situation confers greater rather than lesser or no obligations upon us toward those at our mercy.

In discussing the contact lens experiment with University of California poultry researcher Joy Mench, I mentioned the fact that the protocol did not even contain a requirement for the veterinary care and treatment of the sensitive eyes of living creatures being used in so intimate and destructive a fashion (Davis 1992). Would we treat a human being this way? She said that, after all, we have decided that we can use these animals, so the same ethical standards do not apply.

Could there be a better reason to get the slaughterhouse out of our kitchens and out of our lives? The basic premise of our relationship with the animals we enslave and call "food" invites a low standard of behavior toward them. It encourages us to "decide" that morality does not apply in their case, even though our treatment of our fellow creatures is intrinsically a matter of morality, including the decision to flout morality in order to practice science or produce food. The basic premise governing our relationship to "food" animals engenders guilt and rationalizations, even hatred of our innocent victims, because we know, as Henry Spira observed, that our victim's heartbeat is also our own. He asked, "What gives us the right to violate the bodies and minds of other feeling beings?" We have a sneaking suspicion that we have no right. People who know and like chickens are reduced to deception and denial: "The first rule to remember if you plan on raising chickens for meat is never to name a bird you intend to eat! . . . If you name your future meat, call it Colonel Sanders or Cacciatore. Above all, try to take a lighthearted attitude toward the matter. It's the only way you'll be able to do your own butchering and keep your sanity" (Luttmann and Luttmann, 101).

It is sometimes claimed in contrast to our own culture that traditional cultures believed that by eating a certain kind of animal they incorporated the animal's virtues and spirit. However so, the fact is that in our society, millions of people chomp on dead chickens, dead wings, and leg stumps, and make crude chicken jokes, absurdly constituting themselves by the intimate act of eating beings and products from beings they call "dirty" and whom they despise.

Morally Handicapped Industry

I can assure you that the animal welfare issue (a) will not disappear, and (b) cannot be solved by public relations alone. There is a danger that if this nettle is not grasped, animal agriculture will be seen as ethically challenged or morally handicapped."

—Ian Duncan, "The Science of Animal Well-Being," 1993

Even so, the poultry industry and agribusiness generally worry that the public may come to perceive them as morally handicapped, as indeed they are. It is a sign of moral handicap to mutilate the mouth of a bird, cage her for life, starve her for money, and propose blindness as a "solution" to her suffering. It is a sign of moral handicap to force chickens and turkeys to grow so big so fast that it is painful for them merely to stand on their feet, to take away chickens' feathers, make fun of them, and force them to huddle naked together in their own waste waiting to be killed. The poultry industry is not only cruel, but obscene. It isn't only the masturbation and artificial insemination of "breeder" turkeys and increasingly of chickens, ducks, and geese (Bakst, et al.), or the sticking of balloons and tampons in the uteri of laying hens and making them die a death that only a savage would conceive of. For thousands of years, human beings have violated the bodies and family life of birds and other living beings. We have reached the point where poultry researchers blandly assert, "We are no longer selling broilers, we are selling pieces. A knowledge of how broilers of different strains and sexes grow and become pieces is increasingly important" (Dudley-Cash 1992, 11).

The Future

As the birds have gotten larger, the foodservice market wants smaller pieces.

—Steve Bjerklie, "The Era of Big Bird," January 1, 2008

Walk into any meat or dairy section of your local grocery or natural foods store and you'll notice the labels: "Certified Humane," "Naturally Raised," "Cage-Free," "Organic," "Free-Range" and so on. These give the vague impression that the animals used or killed are given a certain level of consideration, allowed a somewhat natural life.

—Editorial, *Satya*, September 2006

Some people believe that we are moving in the direction of "humane meat" and "animal-friendly" agriculture as people become better informed about the realities of industrialized animal production practices. However, a global decline in industrialized animal farming is not going to happen as long as billions of people are consuming animal products. At the very time that experts are calling farmed-animal production "one of the top two or three most significant contributors to the most serious environmental problems, at every scale from local to global," trends indicate that the global production of animal products could very well double by 2050 (Steinfeld).

The idea of a past Golden Age of compassionate animal farming that could somehow be reclaimed and modernized is misplaced. Books such as Keith Thomas's *Man and the Natural World,* Richard Ryder's *Animal Revolution,* Charles Patterson's *Eternal Treblinka,* and Benjamin and Pierce's *Marketing Poultry Products* give the lie to such wishes. Not only ducks and geese, but chickens and turkeys have traditionally been force-fed in the procedure known as "cramming" or "noodling" in order to increase the size and growth rates of their livers and bodies, to cite one example. A photograph of turkeys being "noodled" appears on page 345 of the March 1930 issue of the *National Geographic,* along with much else that helps to explain why a sixteenth-century British observer wrote of animals raised for food: "They feed in pain, lie in pain, and sleep in pain" (Thomas, 94).

"Designer" Animals

When the chicken genetic map is completed it should provide a means of selecting production traits for a variety of desires.

—Hans Cheng, USDA Agriculture Research Service

Reviewing current trends, avian scientist Lesley Rogers and others see global animal agriculture increasingly focused on creating genetically modified "designer" animals intended to produce certain kinds of meat and other products for the marketplace. Designer animals "will still have minds," says Rogers, but they will not be treated as such given that their higher cognitive abilities are already "ignored and definitely unwanted" (Rogers 1997, 184). An accompanying aim is thus to create animals so mentally blunted that they will accept the worst possible living conditions, on which cheap animal products depend. Geneticist Bill Muir, who breeds chickens to live passively in battery cages, says that "adapting the bird to the system makes more sense" than the other way around: "By selecting for chickens that could tolerate the social stress, we also get chickens that could tolerate environmental stress," he explains (Sigurdson).

Where do we go from here?

An engineer predicts that the future of chicken and egg production will go something like this. "Mature hens will be beheaded and hooked up en masse to industrial-scale versions of the heart-lung machines that brain-dead human beings need a court order to get unplugged from. Since the chickens won't move, cages won't be needed. Nutrients, hormones, and metabolic stimulants will be fed in superabundance into mechanically oxygenated blood to crank up egg production to three per day, maybe five or even ten.

"Since no digestive tract will be needed, it can go when the head goes, along with the heart and lungs and the feathers, too. The naked, headless, gutless chicken will crank out eggs till its ovaries burn out. When a sensor senses that no egg has been dropped within the last four or six hours, the carcass will be released onto a conveyer, chopped, sliced, steamed, and made into soup, burgers, and dog food.

"The apotheosis of egg production will have been reached. It's going to happen. It's probably already in the works" (Burruss).

Depending on who we are, we will laugh or not laugh. When I see our chickens out in the yard sunning themselves, or tripping through the grass in their inimitable way, half running, half flying, or when I watch, as I never tire of watching, their ineffable balletic gestures, poses, poises, grace, and wit displayed in the dramas of their society, even their anger with each other at times, I can hardly believe that what our society and our species is doing to them is real. I contemplate the contemporary and futuristic fate of the hen and compare this with the fascinating and winsome creature I know, of whom it has been tenderly written that she is "rich in comfortable sounds, chirps and chirrs, and, when she is a young pullet, a kind of sweet singing that is full of contentment when she is clustered together with her sisters and brothers in an undifferentiated huddle of peace and well-being waiting for darkness to envelop them" (Smith and Daniel, 334). In anguish, I wish that all the chickens of the world could be safely gathered together beneath her wings, in perfect and dreamless sleep.

References

AATA (The Animal Transportation Association). 2000. New Era of Poultry Transporter Launched: Concepts 2000 Becomes a Reality. *AATA* 21.1: 8.

Ackerman, Diane. 1990. *A Natural History of the Senses*. New York: Vintage Books.

Adams, Carol J., and Marjorie Procter-Smith. 1994. Taking Life or "Taking on Life"? Table Talk and Animals. *Ecofeminism and the Sacred.* New York: Continuum.

Adams, Richard. L. 1992. Effect of Red Plastic Lenses on Egg Production, Feed per Dozen Eggs, and Mortality of Laying Hens. *Journal of Applied Poultry Research* 1: 212-220.

Aiello, Michael R. 2007. Avascular Necrosis, Femoral Head. *eMedicine from WebMD*, 7 June. http://www.emedicine.com/radio/topic70.htm.

Alternatives for Spent Hens. 2003. *Egg Industry* June: 3.

American Proteins: Water is By-Product. 2003. *WATT Poultry USA* April: 38, 40.

Animals' Angels. 2007. Sara Lee Turkey Transport Investigated by Animals' Angels 25 January. http://www.upc-online.org/transport/040607saralee.html.

Anonymous neighbor. 2003. Tunnel Housing. Email to UPC, 31 March.

AP (Associated Press). 2006. Foam Approved to Kill Chickens in a Pandemic. http://msnbc.msn.com/id/15590706, 6 November.

Appleby, Michael C. 1991. *Do Hens Suffer in Battery Cages: A Review of the Scientific Evidence Commissioned by the Athene Trust* (Edinburgh, Scotland: Institute of Ecology and Resource Management, University of Edinburgh, October.

Appleby, Michael C., et al. 1991. *Applied Animal Behaviour: Past, Present and Future: Proceedings of the International Congress Edinburgh 1991*. Universities Federation for Animal Welfare.

Appleby, Michael C., at al. 2004. *Poultry Behaviour and Welfare*. Cambridge, MA: CABI Publishing.

ARI (Animal Rights International), et al. 1994. Has anyone betrayed more animals than the American Veterinary Medical Association? (Advertisement) *The New York Times* 21 June: A17. http://www.upc-online.org/avma/AVMAad.pdf.

Austin, Wayne (Simmons Engineering Company). 1994. Letter to Clare Druce, 1 February.

[The] Avian Brain Nomenclature Consortium. 2005. Avian Brains and a New Understanding of Vertebrate Brain Evolution. *Nature Neuroscience Reviews* 6 (February): 151-167. Avianbrain.org.

Avian Influenza (Bird Flu) – *What You Need to Know: A Report by United Poultry Concerns*. 2007. http://www.upc-online.org/poultry_diseases/birdflu.pdf.

AVMA (American Veterinary Medical Association). *2000 Report of the AVMA Panel on Euthanasia. JAVMA (Journal of the American Veterinary Medical Association)* 218.5 (1 March).

AWA (*Federal Laboratory Animal Welfare Act, Title 7 U.S. Code*, Sections 2131-2156). Detailed Regulations and Enforcement: *Title 9 Code of Federal Regulations*, Parts 1-4. For a discussion of the Animal Welfare Act, see chapter nine in Gary L. Francione, *Animals, Property, and the Law*. 1995. Philadelphia: Temple University Press.

Bakst, M.R. and G. J. Wishart, eds. 1994. *Proceedings: First International Symposium on the Artificial Insemination of Poultry*. Savoy, Illinois: The Poultry Science Association.

Banzett, Robert B. and Shakeeb H. Moosavi. 2001. Dyspnea and Pain: Similarities and Contrasts Between Two Very Unpleasant Sensations. *APS [American Pain Society] Bulletin* 11.2 (March-April). http://www.ampainsoc.org/pub/bulletin/mar01/upda1.htm.

Baskin, Cathryn. 1978. Confessions of a Chicken Farmer. *Country Journal* April: 37-41.

Baumel, Syd. 2006. Indecent Eggsposure: How Eggs are Laid in Canada. *The Aquarian* (Summer). http://www.upc-online.org/canada/60506eggs.html.

Baxter, M[ichael]. R. 1994. The welfare problems of laying hens in battery cages. *The Veterinary Record* 11 June: 614-619.

Beard, Charles W. 2000. Our Industry Perspective on Research Needs for Salmonella enteritidis. FDA-sponsored SE Egg Meeting: *Salmonella* Enteritidis Research, Atlanta, GA, 8 September.

Behrends, Bruce R. 1992. Is feed the problem? *Egg Industry* September-October: 17-18.

Bell, Donald. 1967. The Economics of Various Molting Methods. *Feedstuffs* 1 July: 21, 24, 42.

Bell, Donald. 1990. An egg industry perspective. *Poultry Digest*. January.

Bell, Donald. 1995. Forces that have helped shape the U.S. egg industry: the last 100 years. *The Poultry Tribune* September: 30-43.

Bell, D[onald]. 1996 *Moulting Technologies – Welfare Issues*, October: 1-9.

Bell, Donald D., and William D. Weaver, Jr., eds. 2002. *Commercial Chicken Meat and Egg Production*, 5th ed. Norwell, Mass: Kluwer Academic Publishers.

Bell, Donald D., and Douglas R. Kuney. 1992. Effect of Fasting and Post-Fast Diets on Performance in Molted Flocks. *Journal of Applied Poultry Research* 1:200-206.

Bell, Donald. 2005. An Alternative Molting Procedure. *Egg Industry* August: 13-16.

Bellett, Gerry. 2003. Harvester Corrals Chickens. *Vancouver Sun* 6 June.

Benjamin, Earl W., and Howard C. Pierce. 1937. *Marketing Poultry Products*. New York: John Wiley & Sons.

Berg, Lotta. 2006. Avian influenza. Email to Vivian Leven, 14 November.

Betrays Oath. 1999. American Veterinary Medical Association Betrays Veterinarians' Oath. *Poultry Press* Fall. http://www.upc-online.org/fall99/avma_betrayal.html.

Bilgili, S.F. 1992. Electrical Stunning of Broilers-Basic Concepts and Carcass Quality Implications: A Review. *Journal of Applied Poultry Research* 1.1 (March): 135-146.

Bilgili, S.F. 2008. Contact Dermatitis in Poultry. *WATT Poultry USA* January: 42-46.

Birchall, Annabelle. 1990. Kinder ways to kill. *New Scientist* 19 May: 44-49.

Bjerklie, Steve. 2008. The Era of Big Bird: The Eight-Pound Chicken is Changing Processing and the Industry. *Meat & Poultry* January: 68 ff. http://www.meatpoultry.com/Feature_Stories.asp?ArticleID=90548.

Blake, John P. 2007. In the Beginning, There Should Be Litter Management. *WATT Poultry USA* May: 38-41.

Bock, David. 1991. Animal rights groups address allegation of abuse in Poultry Unit. *Mustang Daily* 6 June.

Bowers, Pamela. 1993a. Look Beyond the Obvious. *Poultry Marketing and Technology* June-July: 16.

Bowers, Pamela. 1993b. A Diagnostic Dilemma. *Poultry Marketing and Technology* August-September: 18-19.

Bowers, Pamela. 1997. Lowering Catching Stress: Automatic broiler harvesters that reduce worker and bird stress are becoming a reality. *Poultry Marketing & Technology* February-March: 16-17.

Boyce, John R. 1992. AVMA. Letter to Karen Davis, 15 December.

Boyd, Freeman. 1994. Humane Slaughter of Poultry: The Case Against the Use of Electrical Stunning Devices. *Journal of Agricultural and Environmental Ethics* 7.2: 221-236.

Brambell, F.W.R. 1965. *Report of the technical committee to enquire into the welfare of animals kept under intensive livestock husbandry systems*. Command Paper No. 28365. HMSO, London.

Brant, A.W., et al. 1982. *Guidelines for Establishing and Operating Broiler Processing Plants*, Agricultural Handbook Number 581, Washington, DC: Agricultural Research Service, U.S. Department of Agriculture, May: 23.

Braunstein, Mark Mathew. 1993. *Radical Vegetarianism*. Quaker Hill, CT: Panacea Press.

Briggs, Joe Bob. 1993. For Chickens, It's Home, Home on the Free Range. *Telegram-Tribune* (San Luis Obispo, CA) 25 March: 24.

British scientists want more hen space. 1990. *Poultry Digest* May: 44.

Broiler Industry. 1976. July.
Broiler Industry. 1996. May.

Bronstein, Scott. 1991. Chicken, How safe? *The Atlanta Journal-Constitution* 26 May: C1-C5.

Brooks, Jane. 1994. This won't hurt a bit. *The News Journal* (Wilmington, DE) 10 January: D1.

Brown, Lindsay. 2007. Feeding the Ideal Bird. *SPARKplug Submissions* (*AnimalNetsynopsis*) 15 June.

Brown, Paul. 2004. Mad-Cow Disease in Cattle and Human Beings. *American Scientist* 92.4 (July-August): 334-341.

Brown, Robert H. 1990. Broiler breeders new competition for spent layer markets. *Feedstuffs* 10 September: 27.

Brown, Robert H. 1992. Spent fowl turning into popular school lunchroom fare. *Feedstuffs* 20 January: 22.

Brown, Robert H. 1997. Renderers begin accepting spent hens. *Feedsuffs* 20 January.

Brown, Rod. 1993. Egg producers concerned about loss of spent fowl slaughter market. *Feedstuffs* 20 December: 1.

Brown, Rod. 1994. UEP says industry's best hope is rendering. *Feedstuffs* 12 December: 17.

Bueckert, Dennis. 2003. If You Are What You Eat, Then Canadian Cattle May Be Pigs Or Chickens. *Canadian Press* 1 June. http://cnews.canoe.ca/CNEWS/Canada/2003/06/01/101175-cp.html.

Burns, Patricia. 2006. Chickens "Contented," Not Happy. *Intelligencer Journal* 24 August. http://local.lancasteronline.com/6/25108.

Burruss, Robert. 1993. The Future of Eggs. *The Baltimore Sun* 29 December: 16A.

Burton, Robert. 1994. *Egg: A Photographic Story of Hatching*. London; New York: Dorling Kindersley.

Butler, Virgil. 2003a. Sadistic Cruelty in the Chicken Slaughterhouse. Affidavit, 30 January. *Poultry Press* 13.1 (Spring): 10-11. http://www.upc-online.org/broiler/022403tysons.htm.

Butler, Virgil. 2003b. Excerpts from Virgil Butler's Press Conference. Little Rock, Arkansas, 19 February. Courtesy of PETA.

Butler, Virgil. 2003c. One-leggers. Email to Cem Akin, 18 April.

Butler, Virgil. 2003d. Email to Karen Davis, 15 June.

Butler, Virgil. 2004. Clarification on Stunner Usage. The Cyberactivist 27 May. http://cyberactivist.blogspot.com/2004/05/clarification-on-stunner-usage.html.

Callicott, J. Baird. 1980. Animal Liberation: A Triangular Affair. *Environmental Ethics* 2: 311-338.

Calnek, B.W. et al., eds. 1991. *Diseases of Poultry*, 9th Edition. Ames: Iowa State University Press.

Campaigners say eggs contain toxic drugs. 2002. *news.bbc.co.uk/hi/English/health/newsid_1764000/1764135.stm*, 16 January. (Report on a Soil Association study.) http://www.soilassociation.org/antibiotics.

Carlile, Fiona S. 1984. Ammonia in Poultry Houses: A Literature Review. *World's Poultry Science Journal* 40: 99-113.

Carlson, John. 2003. Ward Poultry Farm file. County of San Diego Department of Animal Services. 2 Sept. File addressed to United Poultry Concerns 26 September 2003.

Carr, Lewis E. and James L. Nicholson. 1982. Delmarva Broiler Facilities – State of the Art. *Transactions of the ASAE* (American Society of Agricultural Engineers): 740-744.

CEC (Consortium Executive Committee). 1988. *Guide for the Care and Use of Agricultural Animals in Agricultural Research and Teaching*, 1st ed. Washington, DC: National Association of State Universities and Land-Grant Colleges.

Chalmers, Sarah. 2007. The Real Price of the L2 Chicken. *Daily Mail* 12 August.

Chaucer, Geoffrey. 1949.1975-1977. The Nun's Priest's Tale in The Canterbury Tales. *The Portable Chaucer*. Ed. Theodore Morrison, Revised ed. New York: Viking Penguin. First published 1475.

Cheever, Holly. 1993. Letter to Karen Davis, 18 September.

Cheng, Hans H. 1994. The chicken genetic map: a tool for the future. *Poultry Digest* 53.6 (June): 24-28.

Chicken Hell: Research & Investigations Takes You Inside a Hatchery. 1993. *PETA News* Vol. 8, No. 2 (Spring): 10-12.

Chicken-Rage (How to deal with chicken-rage). 2000. *The Economist* 29 April: 79.

Chickens Aren't the Only Ones. 1987. *Reading Rainbow*. Lancit Media Productions for Great Plains National/ Nebraska ETV Network and WNED-TV, Buffalo, NY.

Chick Master: More than 55 years of product innovation, customer support. 1995. *Poultry Tribune* September: 54.

Christen, Kris. 2001. Environmental News. *Environmental Science and Technology* 35.9, 1 May: 184A-185A.

Church, Daniel C. 1995. A question of the chicken or the egg. *The Intelligencer* (Doylestown, PA) 5 September.

Citizens Petition: Forced Molting of Laying Birds. 1998. Food and Drug Administration Docket No. 98P-0203/ CPI. Filed 14 April. http://www.upc-online.org/spring98/govt_molting_petition.html. http://www.upc-online.org/980401moltrpt.html.

CIWF (Compassion in World Farming) Trust. 2004. *Practical Alternatives to Battery Cages For Laying Hens: Case studies from across the European Union. A Report for Compassion in World Farming Trust.* Petersfield, Hampshire: UK. ciwftrust@ciwf.co.uk. www.ciwf.org.

CIWF Trust. 2006. The Way Forward for Europe's Egg Industry: Keeping the Ban on Battery Cages in 2012.

CIWF. 2006a. Don't let the EU "chicken out" on welfare improvements. *Farm Animal Voice*. No. 161 (Spring): 4.

CIWF. 2006b. Talks on EU Chicken Welfare Directive Collapse. Press Release, 11 December.

Clark, Dustan, et al. 2004. Understanding and Control of Gangrenous Dermatitis in Poultry Houses. http://www.thepoultrysite.com/FeaturedArticle/FATopic.asp?Display=173.

Clark, Dustan and Vijay Durairaj. 2007. Necrotic Enteritis. *The Poultry Site*. http://www.thepoultrysite.com/articles/846/necrotic-enteritis. Originally published in *Avian Advice Newsletter* (University of Arkansas) 9.2 Summer 2007.

Clark, Kim. 1992. The dirt, the smell – the pay cut: A tough job just got worse. *The Baltimore Sun* 6 April.

Clarke v Golden Egg Farm PTY, Ltd., C/No. 36539/92. Hobart Magistrates Ct. 1992.

Clifford, John. 2007. Statement by Dr. John Clifford on H5N2 Detection in West Virginia. USDA-APHIS Press Release, 1 April.

Clifton, Merritt. 2000. Starving the hens is "standard." *Animal People* May: 1, 8.

Clifton, Merritt. 2004. How the U.S. kills sick & "spent" chickens. *Animal People* March.

Clouse, Mary. 2003. Tunnel houses. Email to Karen Davis, 15 June.

Coats, C. David. 1989. *Old MacDonald's Factory Farm: The Myth of the Traditional Farm and the Shocking Truth About Animal Suffering in Today's Agribusiness*. New York: Continuum.

COK (Compassion Over Killing). 2001. *Hope for the Hopeless: An Investigation and Rescue at a Battery Egg Facility*. VHS.

COK. 2003. Mad Cow in the United States. 30 December. www.cok.net/lit/madcow.php.

COK. 2007. Pennsylvania Court Finds that Animal Abuse on Egg Factory Farm is Legal. *Compassionate Action* 5 June. http://www.cok.net.

Coleman, George E., Jr. 1976. One Man's Recollections Over 50 Years. *Broiler Industry* July: 48-60.

Coleman, Marilyn. 1995. The Real Impact of Breeders and Hatchery Upon Condemnations. *World Poultry* 11.3: 17, 19.

Commercial Chicken Production. 1989. University of Delaware. VHS.

Community legislation in force: Document 388L0166. 2000. EUR-Lex.

Consumer Reports. 2007. Dirty Chickens. Even "premium" chickens harbor dangerous bacteria. January. http://www.upc-online.org/whatsnew/11507consumerrpts.html.

Coon, Craig. 1994. Telephone interview with Karen Davis, 25 July.

Costello, J.D. 1981. Hen Rehabilitation. *A Place in the Country.* Nantucket, MA: New England Farm and Home Association. No.13. September: 1-2.

Craig, James V. 1981. *Domestic Animal Behavior: Causes and Implications for Animal Care and Management.* Englewood Cliffs, NJ: Prentice-Hall.

Craig, James V., et al. 1992. Beak Trimming Effects on Beak Length and Feed Usage for Growth and Egg Production. *Poultry Science* 71: 1830-1841.

Cristen, Kris. 2001. Environmental News. *Environmental Science and Technology* 35.9. 1 May: 184A-185A.

Cross, H. Russell. 1992. Letter to Karen Davis, 24 August.

Crowe, Susan. 2004. Chickens and antibiotic resistance. *Farm and Dairy* 6 May: A2.

Cruelty and Salmonella Linked. 1993. *Poultry Press* 3.3 (Summer): 10.

Cummings, Terry. 2003. Email to Mary Finelli, 9 December.

Cunningham, Paul. 1992. "Water belly" no longer an emerging disease. *Med-Atlantic Poultry Farmer* 24 Nov: 9.

Curtis, Stan. N.d. Myths and Truths About Modern Poultry Production. Perdue Farms.

CVMA (Canadian Veterinary Medical Association). 2000. Forced Moulting of Chickens. Fax from the Association of Veterinarians for Animal Rights to United Poultry Concerns, 11 November.

Danbury, T.C. et al. 2000. Self-selection of the analgesic drug carprofen by lame broiler chickens. *The Veterinary Record* 146 (11 March): 307-311.

Daniel, Isaura. 2006. The Race for the Perfect Chicken. *Brazilian-Arab News Agency* 29 June.

Danowski, Nancy. 1994. Letter to Karen Davis, 23 July.

Dark chicken bones not a problem. 1994. *Star Tribune* (Minneapolis, MN): 19 June.

Darst, Paul. 2007. Keeping Our Food Supply Safe. *The State Journal-West Virginia Media* 18 April.

Davis, Karen. 1990. Viva, The Chicken Hen. *Between the Species* 6.1: 33-35.

Davis, Karen. 1991. Researching the Heart: An Interview with Eldon Kienholz. *The Animals' Agenda* April: 12-14.

Davis, Karen. 1992. Conversation with Joy Mench. University of Maryland, College Park, MD, 10 August.

Davis, Karen. 1996. *Prisoned Chickens, Poisoned Eggs: An Inside Look at the Modern Poultry Industry.* Summertown, TN: Book Publishing Company.

Davis, Karen, and Nedim Buyukmihci. 1997. Stop starving chickens for profit. *Providence Journal* (RI). 27 October.

Davis, Karen, and Holly Cheever. 1999. American Veterinary Medical Association should have a consistent ethic on animal care. *Chicago Tribune* 31 March, Sec. 1: 23.

Davis, Karen. 2001. *More Than a Meal: The Turkey in History, Myth, Ritual, and Reality.* New York: Lantern Books.

Davis, Karen. 2003a. Will McDonald's policy cure cruelty to chickens? *San Francisco Chronicle* July 14: B7.

Davis, Karen. 2003b. *The Experimental Use of Chickens and Other Birds in Biomedical and Agricultural Research.* http://www.upc-online.org/genetic/experimental.htm.

Davis, Karen. 2004. Open Rescues: Putting a Face on Liberation. *Terrorists or Freedom Fighters? A Reflection on the Liberation of Animals.* Ed. Steven Best and Anthony J. Nocella, II. New York: Lantern Books. 195-212.

Davis, Karen. 2007. Chickens are more evolved than previously believed. *The Buffalo News* 17 July. http://www.upc-online.org/thinking/71707buffalo.html.

Dawkins, Marian Stamp. 1993. *Through Our Eyes Only? The search for animal consciousness.* Oxford; New York: W.H. Freeman and Company.

Day, Elbert J. 1976. Broiler Nutrition: yesterday today tomorrow. *Broiler Industry* July: 140-143.

"Deep Throat Chicken" just keeps on clucking. 1994. *The Local* (Franklin, Kentucky) July: 2.

Delegates act on animal welfare resolutions. 2004. *AVMA Newsbulletin* 30 July. AVMABulletin@avma.org.

Devine, Tom. 1989. The Fox Guarding the Henhouse. *Southern Exposure* 17.2 (Summer): 42.

Dickens, James. 2002. A Tough Old Bird? No More! *Agricultural Research Magazine* December. http://www.ars.usda.gov/is/AR/archive/dec02/bird1202.htm.

Dickenson, Patricia. 2007. Insight into light and reproduction: Researcher delves into why naturally blind chickens perform better. University of Guelph, 17 June. https://www.uoguelph.ca/research/news/articles/2007/June/light_and_reproduction.shtml.

Donn, Jeff. 2007. Associated Press. New weapon in war on bird flu: tiny bubbles. *MSNBC.com*, 10 June. http://www.msnbc.msn.com/id/19036479/page2/.

Drake, Monica. 2007. For a Small Slaughterhouse, A Dash of Local Glory. New York on Film, *The New York Times* 6 May.

Drew, James. 2005. Activists say tape proves cruelty at Ohio egg farm: group targets "animal care" label. *Toledo Blade* 27 February.

Druce, Clare. 1989. *Chicken and Egg: Who Pays the Price?* London: The Merlin Press.

Druce, Clare, ed. 1993. *Hidden Suffering: Notes to Accompany the Videotape*. Farm Animal Welfare Network (UK).

Druce, Clare. 2004. *Minny's Dream*. Cambridge, England: Nightingale Books.

Druce, Clare. 2006. Give them back the sunshine. *Farm Animal Voice* (Compassion in World Farming) No. 162 (Summer): 7.

Dubbert, William H. 1987. Efforts to Control Salmonella in Meat and Poultry. *Third Poultry Symposium Proceedings: Managing for Profit, ed.* Robert E. Moreng. Fort Collins, CO: Colorado State University.

Duckworth, Barbara. 2003. Alberta enlarges layer cages. *The Western Producer* 13 July.

Duckworth, Barbara. 2005. New hen killing method developed. *The Western Producer* 21 April. http://www.upc-online.org/nr/33005co2.htm.

Dudley-Cash, William A. 1992. Latest research findings reported at annual poultry science meeting. *Feedstuffs* 7 September: 11.

Dudley-Cash, William A. 1995. Commercial cage rearing of broilers should not be ignored. *Feedstuffs* 6 March: 11, 19.

Dudley-Cash, William A. 2007. Non-withdrawal molting reviewed. *Feedstuffs* 4 June: 12-34.

Duncan, Ian J.H. 1993. The Science of Animal Well-Being. *Animal Welfare Information Center Newsletter*. Washington, DC: National Agricultural Library 4.1 (January-March).

Duncan, Ian J.H. 1999. Force-Moulting of Laying Hens. *Can-Ag-Fax. Newsletter of the Canadian Farm Animal Care Trust* (CanFACT) 6.2 (Fall): 1, 5.

Duncan, J.H. and Joy Mench. 2000. Does Hunger Hurt? Letter. *Poultry Science* 79.6 (June): 934.

Duncan, Ian J.H. 2001. Animal Welfare Issues in the Poultry Industry: Is There a Lesson to Be Learned? *Journal of Applied Animal Welfare Science* 4.3: 207-221.

Duncan, Ian. J.H. 2006. Vancouver Humane Society Conference, Vancouver, BC, 7 November.

Dunn, Patricia A. 1996. *Eye Disorders of Poultry. PLT [Poultry Litter Treatment] Poultry Health Sentinel* 2.1 (January).

Durham, Sharon. 2002. Chicken Feather Follicles Don't Harbor Bacteria. ARS News Service 25 June. www.ars.usda.gov/is/pr/thelatest.htm.

Durham, Sharon. 2006. Using Comparative Genomics to Manage Marek's Disease. ARS News Service 6 December. http://www.ars.usda.gov/is/pr.

Editorial. 2006. *Satya*, September.

Editors. 1976. *Broiler Industry*, July: 14.

EECC. 1992. *European Report*-467(AC). AWI Quarterly (Animal Welfare Institute, Washington, DC) 41.3 (Summer): 19.

EFSA (European Food Safety Authority-AHAW panel). 2006. *Scientific Report: The welfare aspects of the main systems of stunning and killing applied to commercially farmed deer, goats, rabbits, ostriches, ducks, geese and quail.* Adopted by the AHAW panel on 13 February: 46-47.

http://www.efsa.europa.eu/EFSA/Scientific_Opinion/ahaw_stunning2_report1,0.pdf.

Egg Production. 1987. Compassion in World Farming Fact Sheet, July.

Elanco Products Company (Division of Eli Lilly)). 1990. The Best Method of Cocci Control Has Been in Development For Over 6,000 Years. *Feedstuffs* 6 August: 13.

Eleazer, T.H. 1990. British scientists want more hen space. *Poultry Digest* May: 44.

Elliot, Michael. 1994. Poorly feathered hens eat more food. *Poultry Digest* Aug: 27.

ENS (Environmental News Service). 2004. High Levels of Arsenic Found in Chicken. *Environmental Health Perspectives* 20 January.

Environment: An Alchemy of Fowl. *1991 Annual Report*. University of Georgia College of Agricultural and Environmental Sciences: 14.

Ernst, Ralph. 1993. Effects of temperature and feather cover on hens. *Egg Industry* November-December: 32.

Espenshade, Charlene M. 2007. "Real World" Experience with New Poultry Depopulation Method. *Lancaster Farming* 12 April. http://www.lancasterfarming.com/node/521.

Europe Bans Battery Hen Cages. 1999. *Poultry Press* 9.3 (Fall): 1

Evans, Terry. 1995. IEC gathers in Stockholm. *Egg Industry* October: n.p.

Evans and Page. 2005. U.S. District Court for the Northern District of California, San Francisco Division. 2005. Complaint for Declaratory and Injunctive Relief, 21 November. http://hsus.org/web-files/PDF/HMSA_complaint.pdf.

Farb, JoAnn. 1993a. Letter to Karen Davis, 1 April.

Farb, JoAnn. 1993b. Letter to Karen Davis, 26 Aug.

Farm Sanctuary. 1991. *Humane Slaughter?* VHS.

Farm Sanctuary. 1993. *Sanctuary News* (Fall): 6.

Farmer Automatic of America. http://www.farmerautomatic.com/Broiler-Matic.php.

FAW (Farmed Animal Watch). 2007a. Cruelty Charges Against Poultry Slaughterer. 16.7, 25 May. http://www.upc-online.org/slaughter/052907mfa.html.

FAW (Farmed Animal Watch). 2007b. Bernard Matthews Again Accused of Assaulting Turkeys. 19.7, 26 June.

FAWC (Farm Animal Welfare Council). 1991. *Report on the Welfare of Laying Hens in Colony Systems*. London: MAFF Publications.

FDA-USDA (Food and Drug Administration & U.S. Department of Agriculture) *Egg Safety Public Meeting Transcript of Proceedings*. 2000. Columbus, Ohio: 30 March.

Feed savings could justify beak trimming. 1993. *Poultry Digest* March: 6.

Feedstuffs. 1992. Advertisement 16 March: 27.

Finelli, Mary. 2002. Email to Karen Davis, 11 November.

Fitzsimons. Elizabeth. 2003. No cruelty charges in chicken killings. *San Diego Union-Tribune* 11 April.

Fleishman, Vicky. 1993. Country Fair Farms, Westminster, MD, 25 January.

Fliss, Bob. 1993. Salmonella increase linked to shipping. *Southern Poultry* October: 15.

Forced Moulting of Chickens. 2000. Canadian Veterinary Medical Association (CVMA), 11 November.

Fountain, John W. 1995. Chickens Come First in Her Pecking Order. *The Washington Post*, 31 August: C3.

Fowler, N. 1990. Competitive exclusion – the way forward? *International Hatchery Practice* 5.1.

Fox, Michael W. 1983-1984. Animal Freedom and Well-Being: Want or Need? *Applied Animal Ethology* 11: 205-209.

Francis, Norval E. (USDA Foreign Agricultural Service) 2005. Abolition of Battery Cages in UE-25: Cost Estimated at 354 Million. 31 March. http://www.fas.usda.gov/gainfiles/200504/146119304.pdf.

Francois, Twyla (Canada Head Inspector for Animals' Angels). 2006. Email to Karen Davis, 7 December.

Gage, Mary. 1981. *Praise the Egg*. London: Angus & Robertson Publishers.

Galvin, J.W. N.d. Slaughter of Poultry for disease control purposes. AVIAN INFLUENZA discussion paper.

Gay, Lance. 2004. Activists charge mistreated chickens have weak legs. *Scripps Howard News Service* 20 April.

Geist, Kathy. 1991. Letter. *The Friendly Vegetarian*: Washington, DC: Friends Vegetarian Society of North America. Fall: 3.

General Accounting Office report GAO-02-183. 2002. Mad Cow Disease: Improvements in the Animal Feed Ban and Other Regulatory Areas Would Strengthen U.S. Prevention Efforts. January: 25, 50. http://www.gao.gov/new.items/d02183.pdf.

Gentle, Michael J. 1986. Aetiology of food-related oral lesions in chickens. *Research in Veterinary Science* 40: 219-224.

Gentle, Michael J., et al. 1990. Behavioural evidence for persistent pain following partial beak amputation in chickens. *Applied Animal Behaviour Science* 27: 149-157.

Gentle, Michael J. 1992. Pain in Birds. *Animal Welfare* 1: 235-247. Universities Federation for Animal Welfare.

Germany: Layer Cage Ban Retracted. 2006. *Egg Industry* August: 3.

Gerrits, A.R. 1995. Method of Killing Day-Old Chicks Still Under Discussion. *World Poultry* 11.9: 28.

Glatz, P[hilip].C., ed. 2005. *Beak Trimming*. Nottingham, U.K.: Nottingham University Press.

Goldberg, Ray A. 1976. Broiler Dynamics – past and future. *Broiler Industry* July: 14-22.

GRAIN. 2006. Fowl play: the poultry industry's central role in the bird flu crisis. February. http://www.grain.org/go/birdflu.

Grandin, Temple. 2001. What Would the Public Think? Paper presented at the National Institute of Animal Agriculture 4 April. http://www.grandin.com/welfare/corporation.agents.html.

Grandin, Temple. 2002. Animal Handling & Stunning Conference. American Meat Institute. Kansas City, MO: 21-22 February.

Grandin, Temple. 2005a. Hatching Innovations in Poultry Stunning. *Meat & Poultry* 1 July.

Grandin, Temple. 2005b. *Animals in Translation*. New York: Scribner.

Grandin, Temple. 2007. Audited Facilities Maintain High Standards. *Meat & Poultry* July 1. http://www.meatpoultry.com/news/headline_stories.asp?ArticleID=87099

Greger, Michael. 2003. Mad Cow Disease: Don't Just Switch to Chicken. 26 December. www.veganMD.org.

Greger, Michael. 2006. *Bird Flu: A Virus of Our Own Hatching*. New York: Lantern Books.

Gregory, N.G. 1984. *A Practical Guide to Neck Cutting in Poultry*. Meat Research Institute Memorandum No. 54. Agricultural and Food Research Council. Langford, Bristol, UK. August: 1-8.

Gregory, N.G. 1986. The physiology of electrical stunning and slaughter. *Humane Slaughter of Animals for Food*. Potters Bar: UK: Universities Federation for Animal Welfare. 3-14.

Gregory, N.G. 1993. Letter to Karen Davis, 11 January.

Gregory, N.G and S.D. Austin. 1992. Causes of trauma in broilers arriving dead at poultry processing plants. *The Veterinary Record* 131: 501-503.

Gregory, N.G., and S.B. Wotton. 1990. Effect of Stunning on Spontaneous Physical Activity and Evoked Activity in the Brain. *British Poultry Science* 31: 215-220.

Grimes, William. 2002. *My Fine Feathered Friend*. New York: North Point Press.

Gross, Steve. 2003. Email to Karen Davis, 31 January.

Growing Up With Tyson. 1991. *Tyson Update*, June-July: 17-19.

Guite, Lauren. 2006. Avian flu information on new foam technique for culling. Email to avianflu@foodandwaterwatch.org, 8 December. Regarding the National Carcass Disposal Symposium, 4 December.

Gyles, Roy. 1988. Technological Options for Improving the Nutritional Value of Poultry Products. *Designing Foods: Animal Product Options in the Marketplace*. Washington, DC: National Academy Press: 297-310.

Hardy. Thomas. 1984. *Tess of the d'Urbervilles*. Vol. 1 of *The Works of Thomas Hardy in Prose and Verse, 1912*. New York: AMS Press. First published 1891.

Harkavy, Ward. 2005. A Chicken in Every Pot. *The Village Voice* 10 May.

Harrison, Ruth. 1964. *Animal Machines*. UK: Stuart.

Harrison, Ruth. 1991. The myth of the barn egg. *New Scientist* 30 November: 40-43. http://www.upc-online.org/battery_hens/102105harrison.html.

Hartman, Roland C. 1976. Egg income drops behind broilers. *Poultry Digest* May: 4.

Hatching Good Lessons. Alternatives to School Hatching Projects. N.d. United Poultry Concerns. http://www.upc-online.org/hatching.

Hawthorne. Mark. 2006-2007. Bird Flu Slaughter. *Satya* December-January: 40-42.

Healthy, Productive Management Practices of the U.S. Egg Industry. United Egg Producers brochure.

Heat Deaths. 1995. *Egg Industry* August: 1.

Heath, George E., et al. 1994. A Survey of Stunning Methods Currently Used During Slaughter of Poultry in Commercial Poultry Plants. *Journal of Applied Poultry Research* 3: 297-302.

Hernandez, Chris. 1998, 1999. Molt. Emails to United Poultry Concerns, 26 October, 19 January.

Hernandez, Nelson. 2005. Egg Label Changed After Md. Group Complains. *The Washington Post* 4 October: B3.

HMSA (*Humane Methods of Slaughter Act*), *Title 7 U.S. Code*, Sections 1901-1906. Detailed Regulations and Enforcement: *Title 9 CFR*, Part 313, Sections 313.1-313.90.

HMSA, Title 9 CFR: Animals and Animal Products, Subpart I-Operating Procedures. 381.65c.

Hoerr, Frederic J. 1988. Pathogenesis of ascites. *Poultry Digest* January: 8-12.

Hofstad, M.S., et al., eds. 1984. *Diseases of Poultry*, 8th Edition. Ames: Iowa State University Press.

Holt, Peter S. and Robert E. Porter, Jr. 1992. Effects of induced molting on the course of infection and transmission of *Salmonella enteritidis* in white leghorn hens of different ages. *Poultry Science* 71: 1842-1848.

Holt, Peter S., et al. 1994. Effect of Two Different Molting Procedures on a *Salmonella enteritidis* infection. *Poultry Science* 73: 1267-1275.

Hopey, Don. 2008. Chicken Feed Additive May Pose Danger. *Pittsburgh Post-Gazette* 7 February. http://www.post-gazette.com/pg/08038/855502-114.stm.

Horton, Tom. 2006. 42-Day Wonders. *Washingtonian* September: 66-79.

Hotchkiss, Bruce. 1992. Rural Ramblings: "Chickens are uncooperative." *The Delmarva Farmer* 3 November: 12.

HSUS (The Humane Society of the United States). 2002. Nat'l Organic Standards Board Affirms Animal Welfare Standards for Poultry, 9 May.

HSUS. 2006. USDA Reverses Decades-Old Policy on Farm Animal Transport. Press Release, 28 September.

HSUS. 2007. Bon Appetit Management Co. Encourages Poultry Suppliers to Implement Better Slaughtering Methods. Press Release, 12 July. http://www.hsus.org/press_and_publications/press_releases/bon_appetit_cak.html.

Hubbard, Wentworth. 1976. Exciting future for breeders. *Broiler Industry* July: 30-31.

Huckshorn, Kristin. 1995. The Burden of the Last Muslims. *San Jose Mercury News* 19 May: 1A, Back Page.

Humane Society Asks Tyson to Investigate Freezing Live Birds. 2006. *The Morning News* (Springdale, AR) 10 February.

Hunter, Bruce. 2007. Email to Susan Vickery re: "a rooster neuter," 4 April.

Huntley, John. 1995. UPC Interview with New York State Veterinarian Dr. John Huntley. Phone interview with Karen Davis, 24 May. http://www.upc-online.org/alerts/121404interview.htm.

Hunton, Peter. 1993. Genetics and Breeding as They Affect Flock Health. *The Health of Poultry*. Ed. Mark Pattison. Harlow, Essex: Longman Group UK Ltd.

Immune system has three components. 1995. *Poultry Digest* November.

Industry News. 2004. *Egg Industry* February: 3.

Inside a Turkey Hatchery. 2007. *Compassionate Action: The Magazine of Compassion Over Killing* 20 (Winter-Spring): 4-5.

International Exposition Guide. 2007. Mt. Morris, IL: Watt Publishing Co.

Ivins, Molly. 1991. *Molly Ivins Can't Say That, Can She?* New York, Random House.

Japanese Quail. United Poultry Concerns Brochure. http://www.upc-online.org/quails.

Johnson, Andrew. 1991. *Factory Farming*. Oxford, U.K.: Blackwell.

Johnson, Gordon. 1976. Gordon Johnson Remembers. *Broiler Industry* July: 119-120.

Jones, Tamara. 1999. For the Birds. *The Washington Post* 14 November: F1, F4-F5.

Julian, Richard J., et al. 1992. Effect of poultry by-product meal on pulmonary hypertension, right ventricular failure and ascites in broiler chickens. *Canadian Veterinary Journal* 33 (June).

Jull. M.A. 1927. The Races of Domestic Fowl. *National Geographic* Magazine 51.4 (April): 379-452.

Jull, Morley A. 1930. Fowls of Forest and Stream Tamed by Man. *National Geographic* 57.3 (March): 326-371.

Kalmbach Feeds, Inc. N.d. (Upper Sandusky, Ohio). Egg layer molting program.

Kaufman, Marc. 2000. Cracks in the Egg Industry: Criticism Mounts to End Forced Molting Practice. *The Washington Post* 30 April: A1, A10-A11.

Kaufman, Marc. 2005. Ending Battle with FDA, Bayer Withdraws Poultry Antibiotic. *The Washington Post* 9 September.

Kestenbaum, David. 2001. Engineered Animals. *NPR News, Morning Edition* 4 December: 4-5.

Kienholz, Eldon W. 1990. Letter to Donald J. Barnes, 13 February.

Kiepper, Brian. 2006. Renewed Focus on Wastewater Screening. *WATT Poultry USA* October: 20-22.

Kilman, Scott. 2003. New chicken catcher ruffles fewer feathers. *The Wall Street Journal* 16 June.

Kim, Victoria. 2008. Cruelty charges filed against slaughterhouse boss. *Los Angeles Times* 16 February.

Knierim, U. and A. Gocke. 2003. Effect of Catching Broilers by Hand or Machine on Rates of Injuries and Dead-on-Arrivals. *The Journal of Animal Welfare*. Institute of Animal Hygiene, Animal Welfare and Behaviour of Farm Animals, School of Veterinary Medicine, Hanover, Germany.

Knowles. T.G. 1994. *World's Poultry Science Journal* 50: 60-61.

Knowles, Toby et al. 2008. Leg Disorders in Broiler Chickens: Prevalence, Risk Factors and Prevention. *PLoS ONE* 3.2. 6 February: e1545.

Koene, P., and P.R. Wiepkema. 1992. Pre-dustbathing vocalizations of hens as an indicator of a "need." *Applied Animal Behaviour: Past, Present and Future: Proceedings of the International Congress Edinburgh 1991*. Ed M.C. Appleby, et al. *Universities Federation for Animal Welfare*: 90.

Kotzwinkle, William. 1971; 1976. *Doctor Rat*. New York: Avon Books.

Kreager, Kenton. 1994. Managing the laying bird in hot weather. *International Hatchery Practice* 8.5: 33-38.

Kristensen, H.H., and C.M. Wathes. 2000. Ammonia and poultry welfare: a review. *World's Poultry Science Journal* 56 (September): 235-245.

Kristensen, Helle H., et al. 2000. The preferences of laying hens for different concentrations of atmospheric ammonia. *Applied Animal Behaviour Science* 68: 307-318.

Kristensen, H.H., et al. 2001. Depopulation Systems for Spent Hens – A Preliminary Evaluation in the United Kingdom. *Journal of Applied Poultry Research* 10: 172-177.

Kuenzel, Wayne (University of Maryland poultry researcher). 1993. Telephone interview with Karen Davis, 7 October.

Kuepper, George. 2002. *Organic Farm Certification & the National Organic Program*. http://attra.ncat.org/attra-pub/organcert.html.

Kwantes, James. 1999. A new kind of chicken barn. *Abbotsford News* 4 September: C1.

Lacy, Michael. 1995. Mechanized Catching of Broilers. North American Symposium on Poultry Welfare, Edmonton, Alberta, 13 August.

Leach, Roland M. Jr. 1996. Poultry industry should reconsider if bigger is better. *Feedstuffs* 26 August: 10.

Leeson, Steven. 2007. Poultry Nutrition & Health: Dietary allowances. *Feedstuffs 2007 Reference Issue & Buyers Guide*: 44-53.

Leonard, Rodney E. 1991. Animal Rights Concerns Show Media Savvy, Naiveté. *Nutrition Week*. Community Nutrition Institute, 15 March.

Letterman, Tracie (Executive Director of the American Anti-Vivisection Society). 2006. Email to Karen Davis: 8 August.

Levy, Juliette de Bairacli. 1952. *The Complete Herbal Handbook for Farm and Stable*. London: Faber and Faber.

Lewis, Harry R. 1927. America's Debt to the Hen. *National Geographic* 51.4 (April): 453-467.

Live Day-Old Poultry. *Domestic Mail Manual C022*, Sections3.0-3.1.U.S. Postal Service. *Title 39, U.S. Code*.

Lobb, Richard L. 2002. Letter to Madonna Niles. 18 November.

The Local. 2005. EU to Block Swedish Egg Law, 28 March.

Lorenz, Konrad. 1980. Animals Have Feelings. Trans. E.M. Robinson. *Der Spiegel* 47.

Lundeen, Tim. 2007. Antibiotic bans questioned. *Feedstuffs* 12 March: 1, 3.

Luttmann, Rick and Gail. 1976. *Chickens in your Backyard: A Beginner's Guide*. Emmaus, PA: Rodale Press.

Malone, George. 2003. Ammonia Levels Can Have Effect on Grower Health. *Poultry Times* 22 December.

Mark, Patty. 2001. To Free a Hen. *The Animals' Agenda* 21.4 (July-August):25-26.

Mark, Patty. 2006. Question - 42 day wonder. Email to Karen Davis, 15 October.

Marquis, Michael S. 1999. Freedom of Information Act Minutes of the USDA Farm Animal Well-Being Task Group Meetings, log no. FOIA-99-424. Letter to United Poultry Concerns 7 December.

Martin, David. 1997. Researcher studying growth-induced disease in broilers. *Feedstuffs* May 26: 6.

Martin, Robert. 2008. Fixing the Animal Farms. An Interview with Robert Martin. Brita Belli. *E Magazine* July-August. http://www.emagazine.com/view/?4265.

Mason, Jim, and Peter Singer. 1990. *Animal Factories*. New York: Harmony Books.

Masson, Jeffrey Moussaieff. 2003. *The Pig Who Sang to the Moon*. New York: Ballantine Books.

Mauldin, Joseph. 1999. Noz Bonz. Email to Karen Davis, 24 February.

Mautes, Pia. 2006. Animal Feeds: A Risk Factor in Human Foods? *Meat Processing* (Global edition) 30 November.

McBride, G., et al. 1969. The Social Organization and Behaviour of the Feral Domestic Fowl. *Animal Behaviour Monographs* 2.3: 127-181.

McGreevy, Ronan. 2008. The True Cost of Chicken. *The Irish Times* 22 January.

McGuire, Jennifer. 2003. United Egg Producers Add New Molting Feed Alternatives to Animal Care Guidelines. United Egg Producers Press Release, 17 September. http://www.unitedegg.com/html/news/moltingalternative.html.

Mench, Joy A. 1992. The Welfare of Poultry in Modern Production Systems. *Poultry Science Review* 4: 107-128.

Mench, Joy A. 2001. Email to Karen Davis, 13 July. Faxed Danish study by Gurbakhsh Singh Sanotra. 1999. *Registrering af aktuel benstyrke hos slagtekyllinger*. Dyrenes Beskyttelse.

Merrett, Neil. 2007. Processors called to arms in anti-biotic resistance battle. Report on the 107th General Meeting of the American Society for Microbiology in Toronto, Ontario in May 2007. *Foodproductiondaily.com*/Europe, 5 May.

Miller. Chris. 2001. Cooped Up. *The Vancouver Courier* 27 July: 1, 3, 17.

Miller, Diane. 1994. Queens County Farm Museum, Floral Park, New York. Telephone interview with Karen Davis, 22 December.

Miller, Sharon H. 1996. Increased Marek's Condemnations . . . New Strains or Old Problems? *Broiler Industry* May.

Millman, Suzanne. 2002. Noz Bonz. Email to Pattrice Jones, 22 May.

Mills, D.S., and C.J. Nicol. 1990. Tonic immobility in spent hens after catching and transport. *The Veterinary Record* 126 (3 March): 210-212.

Minimizing Morality. 2004. *Meat News* 16 August.

Morrissey, Christine. 2006. Photo Essay: Paying the Price for "Pampered" Poultry. *Satya* September: 38-39. http://www.free-range-turkey.com.

Morse, Dan. 2006. An Answer to Waste Worries? Disposal by Burning Proposed in Md. For Chicken Farms. *The Washington Post* 23 January: B1.

Muirhead, Sarah. 1993a. Control of heat stress essential to keep hens laying in hot weather. *Feedstuffs* 5 April: 13.

Muirhead, Sarah. 1993b. Multitude of problems plague overweight broiler breeder hens. *Feedstuffs* 5 July: 11.

Muirhead, Sarah. 2007. Report attacks concentrated animal production. *Feedstuffs* 30 July: 4. The Food & Water Watch Report *Turning Farms into Factories* can be found at www.foodandwaterwatch.org.

NAL (National Agricultural Library). 2004. Veterinarians and Biomedical Researchers Agree Animals Feel Pain. *AWIC [Animal Welfare Information Center] Bulletin* 12. 1-2.

NBC (National Broiler Council). N.d. *Careers in the Poultry Industry: A Job is Ready When You Are*. Washington, DC: National Broiler Council with the assistance of Merck Animal Health Division.

Nguyen, HongDao. 2006. Turkey deaths prompt query. *The Mercury News* (San Jose, CA) 22 July.

NIAA (National Institute For Animal Agriculture). *2006-2007 Resolutions and Position Statements*. http://www.animalagriculture.org/aboutNIAA/Resolutions/2006%202007%20Resolution%20and%20Position%20Statements.pdf.

Nicol, Christine, and Marian Stamp Dawkins. 1990. Homes fit for hens. *New Scientist* 17 (March): 46-51.

Nierenberg, Danielle. 2005. *Happier Meals: Rethinking the Global Meat Industry*. Ed. Lisa Mastny. The Worldwatch Institute: Worldwatch Paper 171 (September).

North, Mack O. 1976. Startling changes ahead in production practices. *Broiler Industry* July: 81-98.

North, Mack O., and Donald D. Bell, eds. 1990. *Commercial Chicken Production Manual*, 4th ed. New York: Van Nostrand Reinhold.

Norton, Robert A. 1998. Old Principles and New Research. *Broiler Industry* February: 28-32.

Noske, Barbara. 1989. *Humans and Other Animals: Beyond the Boundaries of Anthropology*. London: Pluto Press.

NTF (National Turkey Federation). 1995. Meat Bird Production/Growout. *Food Safety Best Management Practices for the Production of Turkeys*. December: 10.

Odum, Ted W. 1993. Ascites syndrome: overview and update. *Poultry Digest* January: 14-22.

OIE (International Organization for Animal Health). 2005. *Guidelines for the Humane Killing of Animals for Disease Control Purposes*. January 2005 Terrestrial Animal Health Standards Commission Report. http://www.aphis.usda.gov/NCIE/oie/pdf_files/tahc-guide-hum-kill-jan05.pdf.

O'Keefe, Terrence, 2006. Dakota Provisions: Built to Slice. *WATT Poultry USA* June: 16-23.

O'Keefe, Terrence. 2007. Burger King's Move Puts CAS in Spotlight. *WATT Poultry USA* July: 22-25.

Olejnik, Barbara. 2003. Egg industry addresses Newcastle concerns. *Poultry Times* 17 February.

Olentine, Charles. 2001. Stay Out of the Corner. *Egg Industry* December: 14.

Olentine, Charles. 2002. Welfare and the Egg Industry. *Egg Industry* October: 10-26.

Oppenheimer, Judy. 1995. A "Cutthroat" Business: Walking the line of Halachah at Empire Kosher Poultry. *Baltimore Jewish Times* 2 June: 44-49.

Painter, John Jr. 1993. How did the chickens cross the road? Well-scrambled. *The Oregonian* 19 January: A1, A14.

Pam Clarke: independent spirit challenging convention. 1993. *Animal Liberation Magazine* 45 (July-September): 4.

Patterson, Charles. 2002. *Eternal Treblinka: Our Treatment of Animals and the Holocaust.* New York: Lantern Books.

Peace Corp Information Collection and Exchange. 1981. *Practical Poultry Raising Manual M-11.* Washington, DC: Transcentury Corporation.

Peguri, Alfredo, and Craig Coon. 1993. Effect of Feather Coverage and Temperature on Layer Performance. *Poultry Science* 72: 1318-1329.

PETA (People for the Ethical Treatment of Animals). 2008. Tyson Workers Caught Torturing Birds, Urinating on Slaughter Line. January. http://getactive.peta.org/campaign/tortured_by_tyson.

Petition for Rulemaking Regarding Regulations Issued Under the Poultry Products Inspection Act (PPIA), 21 U.S. Code, Section 451, et seq. 21 November, 1995.

Phelps, Anthony. 1991. Hens fed coarse meals devour profits. *Feedstuffs* 10 June: 11.

Picket, Heather (Compassion in World Farming Trust). 2005. Small Game Birds Question. Email to Karen Davis, 15 March.

Pignews: Chickens and Pigs. 1985-1986. *3-2-1-Contact.* Children's Television Workshop for PBS.

Pope, Albert. 2003. Letter to Karen Davis, 9 February.

Poultry Digest. 1990. June: 4.

Poultry Transport – Unimproved? 1993. *Fact Sheet 34.* Farm Animal Welfare Network, February.

PPIA (Poultry Product Inspection Act). *Title 9 CFR: Animals and Animal Products,* Subpart 1-Operating Procedures. 381.65c.

Preharvest interventions are key to controlling pathogens. 1999. *Food Chemical News* 26 July: 24-26.

Prescott, Matt. 2006. Wing flapping and CAK. Email to Karen Davis, 20 June.

Pressler, Margaret Webb. 1995. "Chicken or Egg?" Today There's No Question Which is First. *The Washington Post* 13 April: D12, D14.

Probert, Debra. 2005. Shocking truth about eggs: An expose of an Ontario battery farm reveals the horrors of Canada's egg industry. *Animal Writes.* The Vancouver Humane Society Newsletter 33 (Autumn): 6-7.

Purvis, Andrew. 2006. Pecking order. *The Guardian* (UK) 4 October.

Raj, Mohan. 2004. Poultry Stunning and Slaughter Seminar. Washington, DC: USDA, 16 December. http://www.upc-online.org/slaughter/10505drraj.htm.

Raj, Mohan. 2005a. Live Poultry Market Video. Email to Karen Davis, 5 February.

Raj, Mohan. 2005b. Small Game Birds Question. Email to Karen Davis, 15 March.

Raj, Mohan. 2005c. Alberta promoting carbon dioxide hen killing. Email to Joyce D'Silva, et al., 26 March.

Raj, Mohan. 2006a. (On behalf of The Humane Society of the United States.) Killing Poultry on Farms During Disease Outbreaks: Current Status and Recommendations.

Raj, Mohan (A.B.M.). 2006b. Recent Developments in Stunning and Slaughter of Poultry. *World's Poultry Science Journal.*

Raj, Mohan. 2007. International Depopulation Standards. Email to Karen Davis, 18 June.

Rathner, Janet. 1998. School egg incubation projects unpopular with animal group. *Potomac Gazette* 13 May: A-10-A11.

Ratzersdorfer, Micheline, et al. N.d. *A Shopping Guide for the Kosher Consumer.*

Raymond, Jennifer. 1990. Letter-essay to Karen Davis, 29 October. How I Learned the Truth About Eggs. 1991. *Poultry Press* 1.2: 1-2.

Reed, Craig A. 1998. USDA-APHIS (Animal & Plant Health Inspection Service), Letter to Karen Davis, 21 August.

Rhorer, Andrew R. 2005. Controlling Avian Flu: The Case for a Low-Path Control Program. *WATT Poultry USA* July: 28-35.

Rice, James E., et al. 1908. *The Molting of Fowls.* Bulletin 258. Ithaca: Cornell University.

Ricker, Hall. 1992. Telephone interview with Karen Davis, 18 September.

Rieger, Steve. 2006. Housing: Stack Deck vs. High-Rise Cage Systems. *Egg Industry* February: 4-7.

Ritson, Joseph. 1802. *An Essay on Abstinence from Animal Food as a Moral Duty.* London: Phillips.

RMAD (Rocky Mountain Animal Defense). 1993. *Raw Footage, Raw Pain: An Inside Look at an Intensive Egg Farm.*

Robbins. John. 1987. *Diet for a New America.* Walpole, NH: Stillpoint Publishing, 69.

Rogers, Lesley J. 1995. *The Development of Brain and Behaviour in the Chicken*: Oxon, UK: CAB International.

Rogers, Lesley J. 1997. *Minds of Their Own: Thinking and Awareness in Animals.* Boulder, CO: Westview Press.

Roland, D.A. et al. 1984. Toxic Shock-Like Syndrome in Hens and Its Relationship to Shell-Less Eggs. *Poultry Science* 63: 791-797.

Russell, Scott and Kevin Keener. 2007. Chlorine: The Misunderstood Pathogen Reduction Tool. *Watt Poultry USA*, May: 22-27.

Ruszler, Paul L, et al. 2004. Stress Determination in Pullets Beak Trimmed at One Day vs. Seven Days vs. No Beak Trimming. Project No. 378. Tucker, GA: U.S. Poultry & Egg Association, November. http://www.poultryegg.org/ResProj/PROJ_378.html

Ryder, Richard D. 1989. *Animal Revolution: Changing Attitudes towards Speciesism.* Cambridge, Mass: Basil Blackwell.

Sadler, Cathy. 1991. Letter. *The Daily Times* (Salisbury, MD) 27 August: 5.

Sad Journeys to Slaughter. 1990. *Agscene* 99 (May-June): 17.

Salmonella Progress Noted But New Pathogen Dangers Seen. 1994. *Food Chemical News* 6 June: 5.

Salmonella testing of raw poultry products: 2005 results. 2006. U.S. Department of Agriculture, Food Safety and Inspection Service 23 February. http://www.fsis.usda.gov/PDF/slides_022306_Lange.pdf.

Salt, Henry S. 1921. *Seventy Years Among Savages.* New York: Thomas Seltzer.

Schleifer, John. 1990. AAAP 32nd annual meeting features skeletal problem talks. *Poultry Digest* October: 10-16.

Schneider, Alison. 2004. Eggs Hold Food Safety Secret. *Thepoultrysite.com*, May 25.

Schorger, A.W. 1966. The Wild Turkey: Its History and Domestication. Norman: *University of Oklahoma Press.*

Schuff, Sally. 2007a. Welfare best practices emerge. *Feedstuffs* 4 June: 5.

Schuff, Sally. 2007b. GIPSA seeks strong poultry contract reforms. *Feedstuffs* 6 August: 1, 4.

Shaffer, David. 2006. Salmonella cases in state raise alarm. Minneapolis *Star Tribune* 3 September: A1, A19.

Shane, Simon. 2005. Future of Gas Stunning. *WATT Poultry USA* April: 16-23.

Shane, Simon M. 2006a. Advances in the Control of Diseases of Egg-Producing Flocks. *Egg Industry* October: 10-11.

Shane, Simon M. 2006b. USDA Approves Foam for Flock Depopulation. *WATT Poultry USA* December: 40-42.

Shane, Simon. 2007. Good Growing Leads to Performance, Quality. *WATT Poultry USA* May: 36-37.

Shear, Michael D. 1995. I-95 Traffic Gets All Fowled Up. *The Washington Post* 25 August: C3.

Shepherd, Lynn. 1993. The Chicken Farm: A Personal Search for a Humane Breakfast. Term paper for a course on Society and Environment, Metropolitan State College of Denver. 5 May: 5-6.

Sigurdson, Chris. 2005. Purdue's "kinder, gentler chicken" moves into real-world test. *Feedstuffs* 16 January: 47.

Simon, Stephanie. 2003. A Killing Floor Chronicle. *Los Angeles Times*, Front Page 8 December.

Singer, Isaac Bashevis. 1979. Forward. *Vegetarianism: A Way of Life*. By Dudley Giehl. New York: Harper & Row.

Singer, Peter, and Jim Mason. 2006. *The Way We Eat: Why Our Food Choices Matter*. USA: Rodale.

Skinner, John L., ed. American Poultry Historical Society. 1974. *American Poultry History 1823-1973: An Anthology Overview of 150 Years*. Madison, WI: American Printing and Publishing, Inc.

Slaughterhouse Worker Turned Activist: UPC Talks with Virgil Butler and Laura Alexander. 2004. *Poultry Press* 14.3 (Fall): 1-4.

Smith, Page, and Charles Daniel. 1975. *The Chicken Book: Being an Inquiry into the Rise and Fall, Use and Abuse, Triumph and Tragedy of Gallus Domesticus*. Boston: Little, Brown. Rpt. 2000. Athens: The University of Georgia Press. All page references are to the first edition.

Smith, Rod. 1995. Cattle industry to continue rolling through large inventory. *Feedstuffs* 6 November: 28.

Smith, Rod. 1997. Huge flock, disposal problems making egg price outlook "even worse." *Feedstuffs* 7 July: 8.

Smith, Rod. 1999. European egg production retreating to past. *Feedstuffs* 5 April: 9, 25.

Smith, Rod. 2003a. UEP research shows new process for inducing molt. *Feedstuffs* 20 October: 1, 13.

Smith, Rod. 2003b. Chicken consumption growth slowing, leading to more consolidation, new focus. *Feedstuffs* 3 November: 8, 14.

Smith, Rod. 2005. UEP to end feed withdrawal. *Feedstuffs* 9 May: 1, 5.

Smith, Rod. 2006. Welfare depends on management. *Feedstuffs* 16 October: 8.

Smith, Rod. 2007. Irradiation rule proposed. *Feedstuffs* 9 April: 1, 4.

Smith, Rod. 2008. EU won't delay ban on cages. *Feedstuffs* 14 January: 5.

Smollett, Tobias. 1966. *The Expedition of Humphry Clinker*. Ed. Lewis M. Knapp. London: Oxford University Press. First published 1771.

Soil Association. 2005. Too hard to crack-eggs with residues-Executive Summary. Information Sheet 2 December. http://www.soilassociation.org/antibiotics.

Solraya.com. 1999. Management Guide in Raising Sasso Chickens. http://www.solraya.com.ph/management_guide.php.

The Sound We Heard Was Singing. 1993. *Poultry Press* 3.1 (Winter): 4-5.

Specter, Michael. 2003. The Extremist. *New Yorker* 14 April: 52-67.

Spent Hen Disposal Across Canada. 2003. *Livestock Welfare Insights* Issue 4, June. http://www.afac.ab.ca/research/species/Articles/spenth.htm.

Spent hens for mink food. 1996. *Egg Industry* December.

Spira, Henry. 1993. Animal Rights: The Frontiers of Compassion. *Peace & Democracy News* 7.1 (Summer): 11-14.

Spurgeon, Devon and Stephen Power. 2001. Lawmakers Put Chicks' Transport Up in the Air. *The Wall Street Journal* 7 November.

Stanley-Branscum, Doll. 1995. Rescuing "Layers." *Poultry Press* 5.2. http://www.upc-online.org/rescuing.html.

Stanley-Branscum, Doll. 1996. Telephone communication to Karen Davis, 15 July.

Stanley, D. 1996. Arthritis from foodborne bacteria? *Agricultural Research* (USDA-Agricultural Research Service) October: 16. See Chicken for Dinner: It's Enough to Make You Sick. http://www.upc-online.org/spring98/chicken_for_dinner.html.

Starving Hens For Profit. 1992. *Poultry Press* 2.3 (Summer).

Steinfeld, H., et al. 2006. Livestock's long shadow. *Livestock, Environment and Development Initiative (LEAD)*. http://www.virtualcenter.org.

Stevens, Jack. 2007. On Chickens. *Animals Voice* May-June: 9.

Stevenson, Peter. 2001. *Animal Welfare Problems in UK Slaughterhouses: A Report By Compassion in World Farming Trust*, July.

Stolfa, Patricia. 1998. USDA-FSIS (Food Safety & Inspection Service). Letter to Karen Davis, 21 August.

Stone, Albert E. ed. 1981. J. Hector St. John de Crevecoeur. *Letters from an American Farmer and Sketches of 18th-Century America*. Middlesex, UK: Penguin Books. First published 1782.

Storey, Samantha. 2004. Know Your Meat. *Seattle Weekly* 7-13 January.

Striffler, Steve. 2005. *Chicken: The Dangerous Transformation of America's Favorite Food*. New Haven: Yale University Press.

Studer, Heinzpeter. 2001. *How Switzerland Got Rid of Battery Cages*, trans. Anja Schmidke. Zurich: Pro Tier International. http://www.upc-online.org/battery_hens/SwissHens.pdf.

Study finds Salmonella in one-third of the EU's egg producers. 2006. *Food Production Daily* 16 June. http://www.foodproductiondaily.com.

Stunning advice: U.K processor Deans Foods takes a new step in its annual welfare policy. 2004. *MeatNews.com*, 1 September.

Sullenberger, David. 1993. Roosters: Who Needs 'Em Anyway? *TimeWarrior Farm Chronicle* (Anthony, NM).

Sullenberger, David. 1994. Letter to Karen Davis, 27 February.

Tenpenny, Sherri J. 2006. *FOWL! Bird Flu: It's Not What You Think*. USA: NMA Media Press.

Terzich, Mac. 1995. Ammonia Kills Flock Performance and Cuts Profits. *PLT [Poultry Litter Treatment] Poultry Health Sentinel*. Veterinary Update from Jones-Hamilton Co. October.

Thomas, Keith. 1983. *Man and the Natural World: A History of the Modern Sensibility*. New York: Pantheon Books.

Thompson, Paul B. 2007. Welfare as an Ethical Issue: Are Blind Chickens the Answer? *Bioethics Symposium: Proactive Approaches to Controversial Welfare and Ethical Concerns in Poultry Science*. USDA/CSREES/PAS. Atlanta, GA: 23 January: 3-5.

Thornberry, F.D., et al. 1975. Debeaking laying stock to control cannibalism. *Poultry Digest* May: 205-207.

Thornton, Gary. 2006. FSIS Wields Stick, Hints at Carrot in Salmonella Initiative. *WATT PoultryUSA* July: 30-36.

Thornton, Lisa. 2007. West Virginia Farm AI Positive. *WATT Poultry USA* May: 4.

Tomlinson, Sylvia. 2003. *Plucked and Burned*. Kearney, NE: Morris Publishing.

Turkey Beak Image. 2006. Email to United Poultry Concerns, 4 August. A Mother Turkey and Her Young. 2007. *Poultry Press* 17.3 (Winter): 2.

Turner, Jacky. 2003. *The Welfare of Broiler Chickens in the European Union*. Compassion in World Farming Trust. http://www.ciwf.co.uk.

TV. 2002. Less Stressed Chickens Mean More Dollars for Poultry Growers. *College Park: The University of Maryland Magazine* 13.3 (Summer): 43.

Twenty-Eight Hour Law, Title 45 U.S. Code, Sections 71-74. Detailed Regulations: *Title 9 CFR*, Part 89.

Ubinas, Helen. 1996. Hatching Responsibility. *The Hartford Courant* 28 January: H4.

UEP (United Egg Producers). N.d. *Recommended Guidelines of Husbandry Practices for Laying Chickens*.

UEP. 1983. Healthy, Productive Management Practices of the U.S. Egg Industry. Decatur, GA.

UEP plans research about induced molting practice. 2000. *Feedstuffs* 7 August: 8.

Ungoed-Thomas, Jon. 2005. "Healthy" Chicken Piles on the Fat. The *Sunday Times* (London) 3 April.

United Egg Producers Animal Husbandry Guidelines for U.S. Egg Laying Flocks. 2000. Updated 2003.

UPC (United Poultry Concerns). 1991. Mourning Vigil Scheduled to Commemorate Chickens Raised for Food. Press Release, 29 March.

UPC. 1992. Red Contact Lenses For Chickens: A Benighted Concept. February.

http://www.upc-online.org/RedLens.html.

UPC. 1998. Advance survey to assess the range and benefits of Induced Molting of layer flocks in the United States. Letter to U.S. Egg Producers, January.

UPC. 2002. *Chickens: Their Dignity, Beauty and Abuse.* VHS/DVD.

UPC. 2003a. *Inside a Live Poultry Market.* VHS/DVD. http://www.upc-online.org/nr/121704livemarket.htm.

UPC. 2003b. *Behavior of Rescued Factory Farmed Chickens in a Sanctuary Setting.* VHS/DVD.

UPC. 2003c. The Animal Welfare and Food Safety Issues Associated with the Forced Molting of Laying Birds. http://www.upc-online.org/molting/52703.htm.

UPC. 2005. Treatment of Live Poultry Before Slaughter 25 October. http://www.upc-online.org/Welfare/102505treatment.html.

UPC. 2006. Mass Depopulation of Poultry as a Disease Control Method 11 July. http://www.upc-online.org/poultry_diseases/71106usda.html.

UPC. 2007. *Avian Influenza (Bird Flu) – What You Need to Know.* http://www.upc-online.org/poultry_diseases/birdflu.pdf.

Upton, John. 2007. Fowl Electricity. *Tracy Press* 30 May.

USDA approves new coccidiosis control product. 2002. *Feedstuffs* 20 December: 18.

USDA. 1982. *People on the Farm: Broiler Growers.* 1982. Washington, DC: Government Printing Office.

USDA-ERS (Economic Research Service). 1993. *Livestock and Poultry: Situation and Outlook,* May.

USDA-FSIS (Food Safety and Inspection Service). 2005. *Treatment of Live Poultry Before Slaughter.* Docket No. 04-037N. Federal Register Notice, 28 September.

USDA-NASS (National Agricultural Statistics Service). 1996a. *Livestock Slaughter 1995 Summary,* March.

USDA-NASS. 1996b. *Poultry Slaughter 1995 Annual Summary,* April.

USDA-NASS. 2005. *Poultry-Production and Value 2004 Summary,* April.

USDA-NASS. 2006. *Poultry Slaughter 2005 Annual Summary,* February.

USDA-NASS. 2007a. *Poultry Slaughter 2006 Annual Summary,* February.

USDA-NASS. 2007b. *Livestock Slaughter 2006 Summary,* March.

Van der Sluis, Wiebe. 1993. Will we ever get rid of the disease? *World Poultry: Special Issue on Coccidiosis* August: 16-18.

Van der Sluis, Wiebe. 2007. Featherless; The Future or an Unsaleable Concept. *WorldPoultry.net.* 4 April. http://www.worldpoultry.net/blogs/id102-14207/featherless_the_future_or_an_unsaleable_concept.html.

Vestergaard, Klaus. 1981a. Aspects of the Normal Behaviour of the Fowl. *Tierhaltung* 12: 1-8.

Vestergaard, Klaus. 1981b. The Wellbeing of the Caged Hen – An Evaluation Based on the Normal Behaviour of Fowls. *Tierhaltung* 12:146-163.

Vestergaard, Klaus.1982. The Significance of Dustbathing for the Well-being of the Domestic Hen. *Tierhaltung* 13: 109-118.

Vestergaard, Klaus. 1987. Alternative Farm Animal Housing: Ethological Considerations. *Scientists Center Newsletter* 9.3: 10-11.

Vestergaard, Klaus. 1993. Feather pecking and chronic fear in groups of red jungle fowl: their relations to dustbathing, rearing environment and social status. *Animal Behaviour* 45: 1127-1140.

Viegas, Jennifer. 2005. Study: Chickens Think About Future. *Discovery News* July 14.

Voiceless: The Fund for Animals. 2007. *From Label to Liable: Scams, Scandals and Secrecy: Lifting the Veil on Animal-Derived Food Product Labelling in Australia.* http://www.voiceless.org.au.

Wabek, Charles. 1987. How stunning affects product quality. *Turkey World* July-August.

Walker, Alice. 1988. Why Did the Balinese Chicken Cross the Road? *Living by the Word: Selected Writings 1973-1987.* New York: Harcourt Brace Jovanovich. 171-172.

Wan, William. 2006. In Pile of Waste, MD. Scientists Dig Up A Response to Bird Flu. *The Washington Post* 20 February.

Warriss, P.D., et al. 1992. Responses of newly hatched chicks to inanition. *The Veterinary Record* 18 January: 49-53.

Wathes, C.M. 1998. Aerial emissions from poultry production. *World's Poultry Science Journal* 54 (September): 241-251.

Wathes, C.M., et al. 2002. Aversion of Pigs and Domestic Fowl to Atmospheric Ammonia. *Transactions of the ASAE* (American Society of Agricultural Engineers) 45.5: 1605-1610.

Wathes, C.M. 2003. Air Hygiene for Broiler Chickens. *In Measuring and Auditing Broiler Welfare*. Ed. C. Weeks and A. Butterworth. Wallingford: CABI Publishing.

Watkins, Susan. 2006. Engineering Clean Water. *WATT Poultry USA* September: 48-50.

Watt, Emily. 2006. Scientists urge ban on sales of fresh chicken. *Sunday Star-Times* 9 July.

WATT Executive Guide To World Poultry Trends 2006/07: The Statistical Reference for Poultry Executives. http://www.wattpoultry.com.

Watts, George. 2007. The Changing American Chicken Industry. *Watt Poultry USA* December: 28-30.

Watts, George, and Connor Kennett. 1995. The Broiler Industry. *The Poultry Tribune* September: 6-18.

Webster, Bruce, et al. 1996. *Update on Hen Disposition*. Paper presented at the 1996 International Poultry Exposition Egg Program (Atlanta, GA) 25 January: 5.

Webster, Bruce. 2002. Animal Handling & Stunning Conference. American Meat Institute. Kansas City, MO, 21-22 February.

Webster, Bruce. 2006. Some Good News About Non-Feed Withdrawal Molting. *Poultry Today* 26 July. http://www.poultryandeggnews.com/poultrytoday/news/20060726/112865.shtml.

Webster, A. Bruce. 2007. The Commercial Egg Industry Should Consider Controlled Atmosphere Stunning for Spent Hens, July. http://www.thepoultrysite.com/articles/864/the-commercial-egg-industry-should-consider-controlled-atmosphere-stunning-for-spent-hens.

Webster, John. 1994. *Animal Welfare: A Cool Eye Towards Eden*. Oxford, UK: Blackwell Science Ltd.

Wegmans Consumer Affairs. 2006. Comments@Wegmans.com 22 March.

Weiss, Rick. 1996. President Orders Overhaul of Meat Safety Inspections. *The Washington Post* 7 July: A1, A10.

Weiss, Rick. 2005. Bird Brains Get Some New Names, And New Respect. *The Washington Post* 1 February: A10. http://www.upc-online.org/alerts/20105post.htm.

[The] Welfare of Battery Hens Regulations 1987 (SI 1987 No. 2020). UK.

Wemelsfelder, F. 1991. Animal boredom: do animals miss being alert and active? *Applied Animal Behaviour: Past, Present and Future: Proceedings of the International Congress Edinburgh 1991*. Ed. M.C. Appleby, et al. UK: Universities Federation for Animal Welfare. 120-123.

Wentink, Henk. 1993. Perspective. *Egg Industry* March/April: 44.

WFPA (World Federation for the Protection of Animals). 1977. *Euthanasia of Dogs and Cats: An Analysis of Experience and Current Knowledge with Recommendations for Research* (April).

Whitlow, L.W. and W.M. Hagler. 2006. Mycotoxins in Feeds. *Feedstuffs 2007 Reference Issue & Buyers Guide* 78.38. 13 September: 77-85.

Williams, Erin. 2006. AOL Hits "Delete" on Battery Cages. HSUS Newswire Press Release 28 April.

Williamson, Elizabeth. 2005. Humane Society to Sue Over Poultry Slaughtering: Suit Demands That Birds Be Killed or Rendered Unconscious Before Butchering. *The Washington Post* 21 November: B2.

Wolfson, David J. 1999. *Beyond The Law*. NY: Farm Sanctuary.

Woolman, John. 1971. *The Journal and Major Essays of John Woolman*. Ed. Phillips P. Moulton. New York: Oxford University Press.

Working Undercover at Perdue. 2004-2005. *The Abolitionist: The Magazine of Compassion Over Killing* 17 (Winter). http://www.upc-online.org/slaughter/11805cokperdue.htm.

Wyatt, Tom. 2006. High Temps Fry 35,000 Chickens. *Post-Tribune* (Merrillville, IN) 3 August.

Yegani, Mojtaba. 2007. Vaccine Reactions in Poultry Flocks. *World Poultry* 16 January. http://tinyurl.com/2c7t3r.

York, Michelle. 2006. Hen Activist Says the War on Cages Will Go On. *The New York Times* 7 May. http://www.nytimes.com/2006/05/07/nyregion/07hens.html.

Young, Emma. 2002. Featherless chicken creates a flap. *NewScientist*.com 21 May.

Young, Tobias. 2006. Recycling chickens: Farmers turn to composting amid collapsed spent-hen market. *The Press Democrat* (Santa Rosa, CA) 22 November.

Zamiska, Nicholas. 2004. FDA Proposes Rule Causing Farms to Cut Salmonella in Eggs. *The Wall Street Journal* 21 September.

INDEX